Design and Analysis of Closed-Loop Supply Chain Networks

Emerging Operations Research Methodologies and Applications
Series Editors: Natarajan Gautam
Texas A&M, College Station, USA
A. Ravi Ravindran
The Pennsylvania State University, University Park, USA

Multiple Objective Analytics for Criminal Justice Systems
Gerald W. Evans

Design and Analysis of Closed-Loop Supply Chain Networks
Subramanian Pazhani

For more information about this series, please visit:
https://www.routledge.com/Emerging-Operations-Research-Methodologies-and-Applications/book-series/CRCEORMA

Design and Analysis of Closed-Loop Supply Chain Networks

Subramanian Pazhani

Advanced Analytics and Optimization Services Group,
SAS Institute

CRC Press
Taylor & Francis Group
Boca Raton London New York

CRC Press is an imprint of the
Taylor & Francis Group, an **informa** business

A CHAPMAN & HALL BOOK

First edition published 2021
by CRC Press
6000 Broken Sound Parkway NW, Suite 300, Boca Raton, FL 33487-2742
and by CRC Press
2 Park Square, Milton Park, Abingdon, Oxon, OX14 4RN

Library of Congress Cataloging-in-Publication Data
Names: Pazhani, Subramanian, author.
Title: Design and analysis of closed-loop supply chain networks /
Subramanian Pazhani, Advanced Analytics and Optimization Services Group,
SAS Institute.
Description: First edition. | Boca Raton : CRC Press, 2021. | Includes index.
Identifiers: LCCN 2020043468 (print) | LCCN 2020043469 (ebook) | ISBN
9780367537494 (hardback) | ISBN 9781003083191 (ebook)
Subjects: LCSH: Business logistics.
Classification: LCC HD38.5 .P39 2021 (print) | LCC HD38.5 (ebook) | DDC
658.7--dc23
LC record available at https://lccn.loc.gov/2020043468
LC ebook record available at https://lccn.loc.gov/2020043469

ISBN: 978-0-367-53749-4 (hbk)
ISBN: 978-0-36753751-7 (pbk)
ISBN: 978-1-003-08319-1 (ebk)

Typeset in Times LT Std
by KnowledgeWorks Global Ltd.

To my daughter Pranikha and my wife Nandhini

Contents

Preface   xi
Acknowledgments   xiii
About the Author   xv

1 Introduction to Closed-Loop Supply Chain Design   **1**
  1.1   Introduction   1
  1.2   Product Flows in Closed-Loop Supply Chain   3
  1.3   Types of Product Returns and Recovery   5
  1.4   Commercial Returns   6
  1.5   Closed-Loop Supply Chain Design with
        Commercial Returns   8
  References   9

**2 Designing Closed-Loop Supply Chain Networks
with Commercial Returns**   **13**
  2.1   Introduction   13
  2.2   Problem Description and Considerations   14
  2.3   The Mixed Integer Linear Programming Model
        (Base Model)   19
  2.4   Analysis of the Illustrative Example   20
  2.5   Effect of Return Parameters on Supply Chain Design
        and Profitability   25
      2.5.1   Effect of Varying Customer Return Rate and
            Customer Acceptance Rate on Total Profit   26
      2.5.2   Variation In Network Design with Varying
            Customer Return Rate and Acceptance Rate   26
      2.5.3   Supplier Selection   27
  2.6   A Realistic Case   28
  2.7   Managerial Implications   30
  2.8   Summary   31
  2.9   Multi-Objective Extensions   31
  References   32

3 Characteristics of Commercial Return Products   **33**
  3.1   Introduction   33
  3.2   Modeling Refurbishing Cost   34
  3.3   Customer Categories and their Perception Toward
        Refurbished Products   36
  3.4   Modeling Low-End Customer Acceptance Rate   38
  3.5   Modeling High-End Customer Acceptance Rate   40
  3.6   Summary   41
  References   43

**4 Network Design Incorporating Quality of Returns
   and Customer Acceptance Behavior**   **45**
  4.1   Introduction   45
  4.2   The Mixed Integer Linear Programming Model
        (Extended Base Model)   46
  4.3   Illustrative Example – Extended Base Model   48
        4.3.1   Comparison between Base Model and
                Extended Base Model Solutions   50
  4.4   Sensitivity Analysis – Extended Base Model   52
        4.4.1   Varying Customer Return Rate, Percentage
                of Low-End and High-End Customers, and
                Discounts Offered for Refurbishing Products   52
        4.4.2   Varying Refurbished Product Quality in the
                Market and Discounts Offered for
                Refurbishing Products   54
                4.4.2.1   Changes in optimal discount levels
                          due to changes in return product
                          quality in the market   55
                4.4.2.2   Changes in optimal network design
                          due to changes in return product
                          quality in the market   56
  4.5   Managerial Implications   58
  4.6   Summary   59
  4.7   Multi-Objective Extensions   60
  References   60

**5 Robust Network Design Incorporating Uncertainties
   in Demand and Return Parameters**   **61**
  5.1   Uncertainty in Closed-Loop Supply Chain   61
        5.1.1   Introduction to Robust Optimization Models   62
  5.2   Problem Description and the Model Framework   65
        5.2.1   The Model   66

5.3	Analysis of the Illustrative Example		67
	5.3.1	Design of the Scenarios for the Illustrative Example	67
	5.3.2	Scenarios Used in the Illustrative Example	68
	5.3.3	Robust Optimization Solution	71
5.4	Analysis of the Robust Optimization Solution		72
	5.4.1	Analysis of the Deterministic Design Solution	73
	5.4.2	Comparison of Deterministic and Robust Designs (Scenarios 1 and 2)	74
5.5	Advantages of Robust Optimization		74
	5.5.1	Scenario with 'Good' Demand – Increasing Return Rate and Acceptance Rate	75
	5.5.2	Scenario with Medium Return Rate and Acceptance Rate – Increasing the Demand	76
5.6	Optimal Discount for the Robust Model		76
	5.6.1	Comparison between the Deterministic and Robust Designs (Optimal Discount)	77
5.7	Summary		78
References			79

Preface

Closed-Loop Supply Chains (CLSC) are the integration of forward and reverse supply chains. In today's business environment with increasing concerns over environmental degradation, legislative compliance, global competition, diminishing supply of raw materials, and consumer demands for ecofriendly products, CLSC networks have become mandatory for firms to operate profitable and stay competitive. Designing CLSCs are relatively complex compared to the forward supply chains, as it integrates both forward and reverse product, information, and financial flows, and the uncertainties associated with them. Particularly, a reverse supply chain is a supply driven flow in which the supply of return products and its quality is highly uncertain and outside the direct control of the company. This book aims to provide insights for companies to design their supply chain network by understanding and incorporating these key return parameters into their design, which directly affects the supply chain profitability.

Designing an integrated forward and reverse supply chain is analyzed with a focus on refurbishing networks with commercial returns. Among the different types of returns in a supply chain, commercial returns are products returned due to customer dissatisfaction, shipping damage, and minor defects. It is one of the major type of returns in the United States (US) and European Union (EU). Driven by liberal return policies at the retailers, commercial returns in the United States are estimated to be greater than 10% of overall in-store sales and approximately 25%–40% of the total online sales. Their return value is estimated to be in hundreds of millions of dollars for a single retailer. Data from the National Retail Federation showed that the United States shoppers returned $396 billion worth of purchases in 2018. According to a 2019 report by Website Builder Expert, 27% of electronic products and 23% of shoes are returned in EU. Returned products are refurbished and/or remanufactured and resold in the market, with function and specification comparable to new products. Given that the commercial returns are inevitable, an efficient supply chain network infrastructure should be in place for profitable business operations. The book focuses on key characteristics of commercial returns including quality of returned products and customer acceptance toward refurbished products. The quality of returned products is incorporated into the network design framework and shown to be an important parameter in deciding optimal discounts on the refurbished products. We see examples of refurbished products in real world like refurbished (renewed) products in Amazon and eBay are categorized as A, B, C, D depending on the returned product quality. Using realistic examples, this book shows the need for incorporating these aspects in designing an efficient CLSC network. The book discusses how and why customer categories and their acceptance behavior is incorporated into the network design. The book also analyzes the interaction of these parameters on supply chain network design and profitability. A modeling framework for incorporating uncertainties in the return product parameters and designing a robust network design is also discussed.

This book will be invaluable for managers in designing a sustainable, robust, and profitable supply chain network. The proposed modeling framework is applied to realistic examples and gives valuable insights to readers to understand the interactions of different parameters on CLSC network design. Ideal for managers, practitioners, and researchers in the area of supply chain network design and optimization.

Acknowledgments

First and foremost, I would like to express my deepest gratitude to my advisor, Prof. A. Ravindran, for encouraging and supporting me throughout my doctoral studies. I feel blessed to have worked under his guidance. I express my deepest gratitude to Prof. T. T. Narendran, who has been extremely supportive both in my research and has guided me in times of difficulties in my personal life. My education would not have turned into reality without his blessings and support. I sincerely thank Prof. A. Ravindran and Prof. Natarajan Gautam for giving me the wonderful opportunity to write this focus book. I also thank all my teachers and mentors for motivating and supporting me to reach my goal.

Life would not have been easy without my friends. There are no words to express my love and affection for them. I thank my friend Madhana Raghavan, for reviewing the chapters and providing valuable suggestions for improving their quality. He has been very patient in correcting all the chapters. I also thank Dr. Abraham Mendoza for reviewing the chapters, and for being such a wonderful and caring friend. His prayers and support helped me get through difficult times. My sincere thanks to all my friends, brothers, and sisters for being such a great emotional support and helping me personally in several aspects. You all stood by me through thick and thin and boosted my confidence to push further to succeed.

I thank Cindy Carelli, Senior Acquisitions Editor at CRC Press, and Erin Harris, Senior Editorial Assistant at CRC Press, for their constant support and encouragement from the inception to the completion of this focus book.

No family is perfect. We argue, we fight, we even stop talking to each other at times. But in the end family is family. I thank my wife, daughter, and parents for their support, love, understanding and encouragement, while I was writing this focus book.

Subramanian Pazhani

About the Author

Subramanian Pazhani is currently working as Senior Operations Research Specialist with the Advanced Analytics and Optimization Services Group, SAS Institute, Cary, North Carolina. He holds a PhD in Industrial Engineering & Operations Research from the Pennsylvania State University. At Penn State, he was the recipient of the Marcus fellowship from the department in Fall 2011. He received his Master's degree in supply chain management from the Indian Institute of Technology, Madras. Prior to this, he obtained his undergraduate degrees in Industrial Engineering from College of Engineering Guindy, Anna University and Indian Institute of Industrial Engineering, Navi Mumbai, India, both in 2009.

His research areas include network design in closed-loop supply chains, inventory optimization in supply chains, inventory routing problems, AGV scheduling in manufacturing and warehousing systems, dispatching and cycle time optimization in semiconductor fabs. His work has been published in prestigious journals such as *International Journal of Production Economics, Computers & Industrial Engineering, International Journal of Production Research, Applied Mathematical Modelling, Applied Soft Computing*, and *International Journal of Operational Research*. He currently serves as an Associate Editor in the *International Journal of Green Computing* and the *International Journal of Enterprise Network Management*.

Introduction to Closed-Loop Supply Chain Design

1

The chapter provides an introduction on the different types of product returns and their recovery options. We will primarily focus on commercial returns, and its characteristics. Commercial returns occur for several reasons, including but not limited to, products returned due to customer dissatisfaction, shipping damages, and minor defects noticed within certain days from product purchase. It is one of the major types of returns in the United States and the European Union (EU). Returns from US shoppers accounted for $396 billion in 2018, according to data from the National Retail Federation (Coresight Research, 2019). This chapter provides a prelude to the modeling frameworks presented in upcoming chapters to design robust and profitable CLSCs.

1.1 INTRODUCTION

In the past two decades, environmental concerns are mounting all over the world due to *global warming*. Although a high consumerism economy has supported the growth of industries, it has been the main contributor to global warming. Waste Electrical and Electronics Equipment (WEEE) is one of the major threats to the environment. In 2006, the United States Environmental Protection Agency (USEPA) predicted that more than 30 million computers will be requiring disposal management in the next few years (Morgan, 2006). Platt and Hyde (1997) claimed that consumers in the United States discarded greater than 12 million computers every year, out of which less than 10%

are remanufactured or recycled, and the remaining are sent to landfills. A decade later, the USEPA conducted a study and realized that the number had increased, but only ever so slightly (from 10% to 15%). Another report from USEPA report (2011) showed that only 27% of electronic waste (e-waste) was recycled. Akcalı et al. (2009) claimed that the landfill capacities in the United States have considerably reduced in recent years and is expected to reduce at a higher rate in the future. According to Knemeyer et al. (2002), the estimated remaining landfill space in 29 states is 10 or more years, 5 to 10 years in 15 states, and less than 5 years in 6 states. Disposed electrical and electronic equipment contain hazardous toxic materials like mercury, lead polybrominated diphenyl ether, hexavalent chromium, and cadmium, cause serious damage to the environment by affecting the ecosystem.

Sustainability of the environment can be improved by manufacturing supply chains adopting effective recycling practices. *Sustainability is defined as meeting the needs of the present generation without compromising the ability of future generations to meet their needs* (World Commission on Environment and Development: Our Common Future, 1987). Recycling practices will reduce waste disposal and benefit the society as a whole (Visich et al., 2005). Businesses themselves might not be often inclined toward addressing environmental concerns. However, governmental legislations like the ones passed by the EU (2000, 2003) along with reduced supply of raw materials have compelled firms to adopt sustainable practices. In the United States, laws such as the Electronic Equipment Recycling and Reuse (2010) Act and Covered Device Recycling Act (2010) mandate manufacturers to collect and recycle used electronic equipment. These laws, along with customer expectations, are holding companies accountable for the impact of their operations and products on the society.

Large companies have started to realize the benefits of using the remaining value from reusable goods. For instance, Apple's *'reuse and recycle program'* collects and recycles their products, including the iPhone, iPad, and Mac. Best Buy efficiently recycles used electronic equipment through its recycling program. Best Buy recycled approximately 80 million pounds of e-waste in 2012. E-wastes contain raw materials such as plastics, aluminum, and steel, which can be used in new product production. Alternatively, they can be used in production of other products. Recycled plastics can be used to make compact benches and playground equipment. Aluminum can be used in automotive parts, ladders, furniture, etc. Steel can be used in making electrical appliances, construction of bridges, automobiles, etc. Circuit boards are sources of precious metals like gold, silver, and copper. Japanese firms are actively reclaiming rare minerals, including gold and silver, from used electronics (NY Times, 2010). Used batteries contain cobalt and iron which can be used to produce new batteries. In summary, recycling of products can:

(1) save natural resources by recycling raw materials instead of virgin materials and thereby reducing the consumption of minerals; (2) conserve energy as reprocessing consumes lesser energy than manufacturing new products; (3) reduce stress on landfills; and (4) comply with governmental legislation and enhance their contribution to corporate social responsibility. One key benefit that motivates companies to adopt recycling practices is reduction in cost of raw materials, which directly contributes to the profitability of the business. Throughout this book, we will be focusing on the profitability aspects of adopting CLSCs.

The remainder of the chapter is organized as follows: Section 1.2 discusses the different product flows in a CLSC system and the need to manage them in an integrated manner. Section 1.3 gives an overview of the types of product returns and recovery options. Section 1.4 discusses the characteristics specific to commercial returns. Section 1.5 provides an overview of CLSC design with commercial returns.

1.2 PRODUCT FLOWS IN CLOSED-LOOP SUPPLY CHAIN

A supply chain consists of all the parties involved, directly or indirectly, in fulfilling a customer request (Chopra and Meindl, 2004). Lee and Billington (1993) define supply chain as a network of facilities that perform the functions of procurement of materials, transformation of materials to intermediate and finished products, and distribution of finished products to customers. The overall goal of a supply chain is to fulfill the end customer's needs and expectations in a cost-efficient manner.

A *forward supply chain* is a collection of entities that produces and distributes new products from the suppliers to the customers. The supply chain typically includes manufacturing plants, suppliers, transport carriers, distribution/warehousing facilities, retailers, and customers themselves. Recycling practices, to reduce waste and improve sustainability, has caused material flow back from the users to the manufacturers. The management of material flow in the opposite direction of the forward supply chain flow is defined as *reverse supply chain* (Stock, 1992). Inefficient handling of returned products can lead to increase the cost of the final product (Mutha and Pokharel, 2009). A *reverse supply chain* is a collection of entities and activities required to collect used products from consumers and reprocess them to recover leftover value or dispose them (Pochampally *et al.*, 2008). The used products from the customers are collected and distributed to the upstream supply chain (recycling centers, manufacturers) for possible recycling and reuse. A typical reverse supply chain consists of three entities: (i) collection centers, where customers return used products; (ii) recovery centers, where inspection and reprocessing of used products is done; and (iii) demand centers for reprocessed products. Reverse supply chain activities includes (Guide *et al.*, 2003): (1) product acquisition to obtain the products from the end-users; (2) logistics to move the products from the points of use to a point(s) of disposal; (3) testing, sorting, and disposal to determine the product's condition and the most economically attractive reuse option; (4) refurbishing to enable the most economically attractive option among repair, remanufacture, direct reuse, recycle, and disposal; and (5) remarketing to create and exploit markets for refurbished goods and distribute them.

A CLSC is the integration of forward and reverse supply chains. However, the integration is challenging due to the differences between forward and reverse supply chains (Visich *et al.*, 2005). A forward supply chain has a demand-driven flow where products move from few entities (manufacturers) to many entities (customers). A reverse supply chain has a supply-driven flow where products move from many entities (customers) to few entities (refurbishing plants). The supply of return products is outside the direct control of the business. Also, the product quantity and quality are uncertain, adding complexity to the design of reverse supply chain (Thierry *et al.*, 1995).

Independent operations and management of both supply chains (forward and reverse) can lead to reduced profitability of the entire supply chain. Many studies (Fleischmann *et al.*, 2001; Pishvaee *et al.*, 2010; Uster *et al.*, 2007) have discussed the importance of integrated modeling and analysis of forward and reverse supply chains. Guide and Wassenhove (2009) defined CLSC management as '*the design, control, and operation of a system to maximize value creation over the entire life cycle of a product with dynamic recovery of value from different types and volumes of returns over time*'. The concept of CLSC is very familiar in the soft drink bottling industries in several parts of the world where empty glass bottles are invariably recycled, requiring logistical arrangements for collecting empty bottles. Today, it has extended to several other industries such as Xerox Corporation, IBM, Hewlett-Packard, Kodak, and Caterpillar, to name a few. Xerox Corporation had a savings of around $200 million by remanufacturing copiers returned at the expiration of their lease contracts (Ferguson, 2009). Caterpillar Inc.'s remanufacturing division had over $2 billion in sales and was the fastest growing division out of all of its other divisions (Ferguson, 2009). IBM collected over one million units of used information technology equipment and made revenue of billions of dollars in the second-hand equipment, parts, and materials markets (Fleischmann *et al.*, 2003). In India, some notable CLSC initiatives have been observed in automobile industries. These include recycling of batteries and tires (Kannan *et al.*, 2009). Companies that produce household and industrial appliances using polymers also are seen to take such initiatives. The adoption of CLSC's are largely driven by its profitable business propositions (Flapper *et al.*, 2005; Guide and Wassenhove, 2009). Many research and empirical studies have shown that practicing CLSC is key for improved profitability and competitive advantage in the market (Flapper *et al.*, 2005; Fleischmann *et al.*, 2001; Guide and Wassenhove, 2009; Visich *et al.*, 2005).

1.3 TYPES OF PRODUCT RETURNS AND RECOVERY

Guide and Wassenhove (2009) categorizes product returns in a CLSC into *Commercial returns*, *End-of-Use (EOU) returns*, and *End-of-Life (EOL) returns*. *Commercial returns* are products returned by customers within a 30-, 60-, or 90-day period after the initial purchase. A consumer can return the product for various reasons such as the product quality not meeting his expectation, incompatible product performance with his needs, shipping damages, finding a better price for the same product, or feeling remorse. *EOU* returns are product returns due to a technological upgrade of product functions, warranty returns, and end of lease returns. *EOL* returns are products that become technically obsolete or no longer contain any utility for the current user. Managing product returns is challenging for firms because there is hardly any information on processing and disposal options, return rates, and unauthorized returns.

Product recovery for *EOU* and *EOL* returns primarily occurs in two ways: (1) the returned products or components of the product can be remanufactured and sold in the secondary markets, for example, remanufactured engines, automobile spare parts, tools, etc. (2) The returned products could be dismantled, and usable components could be refitted onto a new product, for example, lead recycled from used batteries can be added to virgin lead while manufacturing new batteries. The former is most suitable for *EOU* returns and the latter, for *EOL* returns. *EOU* returns need extensive remanufacturing processes. These returns have high variability in the use and thus product disposition and remanufacturing options can vary significantly. *EOL* returns are only used for parts recovery and functional parts could be reused/refitted onto new products. Both the US and European markets have some directives or practices for *EOU* and *EOL* returns. In European markets, legislative norms such as EOL Vehicle Directive (EU, 2000) and the WEEE (EU, 2003) have mandated industries to explore the reuse and recycling options primarily to avoid landfill (Salema *et al.*, 2010; Kannan *et al.*, 2010). Paper recycling directive, *EOL* vehicle directive and the WEEE, enforced by the European Union helped recycling and landfill avoidance. In the European Union, automobiles have the highest recycling percentage (94%), followed by aluminum beverage cans (61.1%) and newspapers (57.9%) (Bellmann and Khare, 1999). In the United States, recycling *EOU* and *EOL* returns is termed as remanufacturing. *Remanufacturing* is an industrial process in which worn-out/broken/used products, known as cores, are restored to useful life (Ostlin *et al.*, 2008). Remanufacturing has existed for centuries, typically for high-value and low-volume items, such as locomotive engines, aircraft (aircraft weapons systems and aircraft carriers), largely funded by the US Military (Guide and Wassenhove, 2009). The remanufacturing sector is larger than the United States' domestic steel industry in terms of sales and employment, with annual sales in excess of $53 billion and over 70,000 remanufacturing firms (Lund, 1996). There are about 2,000 to 9,000 remanufacturing firms in the United States, out of which only 6% were Original Equipment Manufacturers (OEMs) (Hauser and Lund, 2008). Kodak and Xerox have been successful in remanufacturing practices for single-use cameras and refillable toner cartridges (Uster *et al.*, 2007).

Commercial returns is one of the major types of returns in both the United States and European markets. A customer buys a laptop at a retailer and after two weeks of usage finds it to be slow for his needs. He returns the laptop at the retailer and buys a different product. This is an example of commercial returns. The biggest drivers for processing commercial returns are: (1) *sustainability*, reducing waste by recycling used products, thereby reducing the usage of raw materials (2) *profitability*, by reusing the usable value in the products and reduction in raw materials purchasing cost. Commercial returns are refurbished and resold in the market, with function and specification comparable to new products. We will focus on the characteristics specific to commercial returns. Interested readers can refer to Pazhani (2014) for additional reading on type of product returns and recovery.

1.4 COMMERCIAL RETURNS

Enormous amounts of commercial returns happen after the consumers are not satisfied with their purchase or potential consumers/critics have completed their evaluation of the product (Vorasayan and Ryan, 2006). A report by Accenture, a technology consulting and outsourcing firm, says that the costs of consumer electronics returns in 2007 was $13.8 billion in the United States alone. The return rates varied between 11% and 20%, depending on the type of product (Douthit et al., 2011). Approximately 58% of consumer electronics retailers and 43% of OEMs would experience higher return rates than in previous years (Douthit et al., 2011). Driven by liberal return policies at the retailers, commercial returns in the United States are estimated to be greater than 10% of overall in-store sales (Guide and Wassenhove, 2009) and approximately 25%–40% of the total online sales. Their return value is estimated to be in hundreds of millions of dollars for a single retailer. The annual estimates of commercial returns (returned within 90 days of sale) in the United States are in excess of $100 billion (Guide *et al.*, 2006). The electronic industry spends approximately $14 billion annually on managing product returns. Data from the National Retail Federation showed that US shoppers returned $396 billion worth of purchases in 2018 (Coresight Research, 2019). According to the Website Builder Expert report (2019) (based on data from October 2017 to September 2018), 27% of electronic products and 23% of shoes are returned in the EU. Among the countries in the EU, German consumers returned approximately 52% of their online purchases.

Can the businesses restrict customer returns? While some companies see these product returns as a roadblock to achieving profits, a few other companies have adopted strict return policies to restrict the returns, hoping to increase their profit margins. According to a research conducted by the Consumer Electronics Manufacturing Association, a store's return policy is very important for 70% of consumers in their decision to shop there (Pinkerton, 1997). In an empirical study by Jupiter Media Metrix, 42% of online shoppers said they would buy more from the Internet if the returns process was easier (Rosen, 2001). A study by Petersen and Kumar (2010) shows that companies with liberal return policies could maximize their future profits. The authors conducted an empirical study with an apparel company and analyzed their profits with strict and lenient return policies. The observations from the study shows that lenient return policies boost the company's future profits. Based on this study, it is also observed that products return cannot be curtailed and managing these returns efficiently can yield significant profits.

Once the product has been returned, an accurate return classification should be carried out for efficient disposition and to speed up the recycling process (Visich *et al.*, 2005). Douthit et al. (2011) shows that nearly 68% of returns are products that work properly but do not meet customers' expectations for certain reasons and 27% of returns are due to buyer's remorse (situations where customers simply changed their minds). Ferguson *et al.* (2006) classify them as false failure returns – the returns of effectively new products that 'have no verifiable functional defect'. Ovchinnikov (2011) empirically stated that, for an electronic product company, only 5% of product returns were attributed to actual defects in the product. These defects may be small cosmetic blemishes, observable cosmetic blemishes, shipping damages to the containers, or malfunction of hardware/software. However, reverse supply chain infrastructure should be in place to efficiently recycle and reuse the value in these products. The refurbished products are then sold in the market for a discounted price. Today, almost every electronic retailer, including Amazon, Best Buy, and Walmart, has an option for buying refurbished items on their websites. Some everyday examples of industries involved in refurbishing/remanufacturing include electronic products such as mobiles, laptops, computers, televisions, and automotive parts like batteries, tires, and starter motors.

The main challenges with refurbishing commercial returns are: (1) cost for refurbishing, which depends on the condition of the returned product and (2) market for refurbished products, which depends on the perception of the consumers on refurbished product quality. A manufacturer will refurbish the returns only if it is profitable to do so. Refurbishing of returns incur two major cost components: inspection and refurbishing cost. Consumers' willingness to buy the refurbished products depends on the perceived quality of the products for its price (Hazen et al., 2012; Ovchinnikov, 2011; Vorasayan and Ryan, 2006).

1.5 CLOSED-LOOP SUPPLY CHAIN DESIGN WITH COMMERCIAL RETURNS

CLSC and its management have become mandatory for companies to stay competitive and profitable, with product return values estimated to be in hundreds of millions of dollars for a single retailer. A strategic issue in supply chain management is the configuration of the supply chain network that is seen to have a significant effect on the supply chain performance indicators (Pishvaee et al., 2009). Network design complexity in the supply chains have increased significantly due to addition of entities, return products, and the uncertainties associated with them. Given that the commercial returns are inevitable, an efficient supply chain network design (or network infrastructure) should be in place for profitable business operations. Unlike end-of-use and end-of-life returns, commercial returns happen within a short period of time from the initial purchase.

Although aspects of managing CLSCs have been addressed in the literature, there have been no insights into incorporating the two key characteristics of commercial returns: quality of the returned products and customer acceptance toward refurbished products. Given the strategic nature of the problem, ignoring these important return aspects, particularly for commercial returns, will lead to inefficient operations of the entire supply chain, as it directly affects warehouse selection, distribution network plan for returned products, and supplier selection decisions. The aim of this book is to provide insights for companies to design their supply chain network by understanding and incorporating these key return parameters in their design, which directly affect the supply chain profitability. The forthcoming chapters in this book address the planning of CLSCs and develop mathematical models and solution methodologies for designing CLSC networks by integrating both the forward and reverse supply chains.

The framework proposed in this book will help practitioners and readers to answer the questions below: what is the gain in adopting a CLSC with recycling?; how sensitive is your supply chain design and profit to the return parameters?; is it profitable to design your CLSC incorporating these parameters?; what is your supply chains' refurbishing strategy, given return parameters?; and how to incorporate uncertainties of return parameters in the supply chain design phase? The proposed modeling framework is applied to realistic examples and provides valuable insights to practitioners and readers to understand the interactions of different parameters on CLSC network design. The modeling frameworks, provided in this book, can be easily extended to adopt any CLSC network in designing a sustainable, robust, and profitable supply chain network.

REFERENCES

Akcalı, E., Cetinkaya, S., Uster, H. (2009), 'Network design for reverse and closed-loop supply chains: An annotated bibliography of models and solution approaches', *Networks* **53**(3), 231–248.

Bellmann, K., Khare, A. (1999), 'European response to issues in recycling car plastics', *Technovation* **19**(12), 721–734.

Chopra, S., Meindl, P. (2004), *Supply Chain Management: Strategy, Planning, and Operation*, Prentice Hall of India, New Delhi.

Coresight Research (2019), Innovator Intelligence: Solving Retail's Most Expensive Problem With AI-Powered Returns Reduction, Retrieved from http://www.newmine.com/wp-content/uploads/2019/07/Innovator-Intelligence-Solving-Retail%E2%80%99s-Most-Expensive-Problem-with-AI-Powered-Returns-Reduction-Jul-1-2019.pdf

Covered Device Recycling Act (2010), Retrieved from http://www.legis.state.pa.us/WU01/LI/LI/US/HTM/2010/0/0108.HTM

Douthit, D., Flach, M., & Agarwal, V. (2011). A returning problem: reducing the quantity and cost of product returns in consumer electronics. Accenture Communications & High tech, Accenture, 3–9.

Electronic Equipment Recycling and Reuse (2010), Retrieved from http://www.dec.ny.gov/docs/materials_minerals_pdf/ewastelaw2.pdf

EU (2000), 'Directive 2000/53/EC of the European parliament and of the council of 18 September 2000 on end-of life vehicles', *Official Journal of the European Communities*. Retrieved from https://eurlex.europa.eu/legal-content/EN/ALL/?uri=CELEX%3A32000L0053.

EU (2003), 'Directive 2002/96/EC of the European parliament and the council of 27 January 2003 on waste electrical and electronic equipment (WEEE)', *Official Journal of the European Union*. Retrieved from https://eur-lex.europa.eu/legal-content/EN/TXT/?uri=CELEX%3A32002L0096

Ferguson, M. E. (2009), 'Strategic and tactical aspects of closed-loop supply chains, foundations and trends in technology', *Information and Operations Management* 3(2), 103–200.

Ferguson, M., Guide Jr., V. D. R., Souza, G. (2006), 'Supply chain coordination for false failure returns', *Manufacturing Service Operations Management* 8(4), 376–393.

Flapper, S. D. P., Nunen, Jo., Wassenhove, L. N. (2005), *Managing Closed-loop Supply Chains*, Springer, New York.

Fleischmann, M., Beullens, P., Bloemhof-Ruwaardz, J. M., Wassenhove, L. N. (2001), 'The impact of product recovery on logistics network design', *Production and Operations Management* 10(2), 156–173.

Fleischmann, M., van Nunen, J. A. E. E., Grave, B. (2003), 'Integrating closed-loop supply chains and spare-parts management at IBM', *Interfaces* 33(6), 44–56.

Guide Jr. V. D. R., Harrison, T. P., Wassenhove, L. N. (2003), 'The challenge of closed-loop supply chains', *Interfaces* 33(6), 3–6.

Guide Jr. V. D. R., Souza G. C., Wassenhove, L. N., Blackburn, J. D. (2006), 'Time value of commercial product returns', *Management Science* 52(8), 1200–1214.

Guide Jr. V. D. R., Wassenhove, L. N. (2009), 'The evolution of closed-loop supply chain research', *Operations Research* 57(1), 10–18.

Hauser, W., Lund. R. (2008), *Remanufacturing: Operating Practices and Strategies*, Boston University, Boston, MA.

Hazen, B. T., Overstreet, R. E., Jones-Farmer, L. A., Field, H. S. (2012), 'The role of ambiguity tolerance in consumer perception of remanufactured products', *International Journal of Production Economics* 135(2), 781–790.

Kannan, G., Noorul Haq, A., Devika, M. (2009), 'Analysis of closed-loop supply chain using genetic algorithm and particle swarm optimisation', *International Journal of Production Research* 47(5), 1175–1200.

Kannan, G., Sasikumar, P., Devika, M. (2010), 'A genetic algorithm approach for solving a closed-loop supply chain model: A case of battery recycling', *Applied Mathematical Modelling* 34(3), 655–670.

Knemeyer, A. M., Ponzurick, G. T., Logar, M. C. (2002), 'A qualitative examination of factors affecting reverse logistics systems for end-of-life computers', *International Journal of Physical Distribution and Logistics Management* 32(6), 455–479.

Lee, H. L., Billington. C. (1993), 'Material management in decentralized supply chains', *Operations Research* 41(5), 835–847.

Morgan, R. (2006), 'Tips and Tricks for Recycling Old Computers', *SmartBiz*. Retrieved from http://www.smartbiz.com/article/articleprint/1525/-1/58 (August 21).

Mutha, A., Pokharel, S. (2009), 'Strategic network design for reverse logistics and remanufacturing using new and old product modules', *Computers & Industrial Engineering* **56**(1), 334–346.

NY Times (2010). Japan recycles minerals from used electronics. Retrieved from: http://www.nytimes.com/2010/10/05/business/global/05recycle.html? pagewanted=all&_r=0 **(October 4)**.

Ostlin, J., Sundin, E., Bjorkman, M. (2008), 'Importance of closed-loop supply chain relationships for product remanufacturing', *International Journal of Production Economics* **115**(2), 336–348.

Ovchinnikov, A. (2011), 'Revenue and cost management for remanufactured products', *Production and Operations Management* **20**(6), 824–840.

Pazhani (2014). Design of Closed-Loop Supply Chain Networks, The Pennsylvania State University, https://etda.libraries.psu.edu/files/final_submissions/9801

Petersen, J. A., Kumar, V. (2010), 'Can product returns make you money?', *Sloan Management Review* **51**(3), 85–89.

Pinkerton, J. (1997), 'Getting religion about returns', *Dealerscope Consumer Electronics Marketplace* **39**(11), 19.

Pishvaee, M. S., Farahani, R. Z., Dullaert, W. (2010), 'A memetic algorithm for bi-objective integrated forward/reverse logistics network design', *Computers & Operations Research* **37**(6), 1100–1112.

Pishvaee, M. S., Jolai, F., Razmi, J. (2009), 'A stochastic optimization model for integrated forward/reverse logistics network design', *Journal of Manufacturing Systems* **28**(4), 107–114.

Platt, B., Hyde, J. (1997), *Plug into Electronics Reuse*, 13–38. Institute of Local Self Reliance, Washington, DC.

Pochampally, K. K., Nukala, S., Gupta, S. M. (2008), *Strategic Planning Models for Reverse and Closed-Loop Supply Chains*, CRC Press, Boca Raton, FL.

Rosen, C. (2001). 'Ready for the Returns?' *Informationweek*. January 8, 22–23.

Salema, M. I. G., Povoa, A. P. B., Novais, A. Q. (2010), 'Simultaneous design and planning of supply chains with reverse flows: A generic modelling framework', *European Journal of Operational Research* **203**(2), 336–349.

Stock, J. R. (1992), *Reverse Logistics*, Council of Logistics Management, Oak Brook, IL.

Thierry, M., Salomon, M., Van Nunen, J. A. E. E., Wassenhove, L. N. (1995), 'Strategic issues in product recovery management', *California Management Review* **37**(2), 114–135.

USEPA report (2011), 'Facts and Figures on E-Waste and Recycling', Retrieved from http://www.electronicstakeback.com/wp-content/uploads/Facts_and_Figures_ on_EWaste_and_Recycling.pdf

Uster, H., Easwaran, G., Akçali, E., Cetinkaya, S. (2007), 'Benders decomposition with alternative multiple cuts for a multi-product closed-loop supply chain network design model', *Naval Research Logistics* **54**(8), 890–907.

Visich, J., Li, S., Khumawala, B. (2005), 'A framework for the implementation of radio frequency identification technology in closed-loop supply chains: impact and challenges' Innovation Monograph I – Enterprise Resource

Planning: Teaching and Research, SAP University Alliance Program and the Kelley School of Business, Indiana University, Bloomington, IN, November, 7–34.

Vorasayan, J., Ryan, S. M. (2006), 'Optimal price and quantity of refurbished products', *Production and Operations Management* **15**(3), 369–383.

Website Builder Expert report (2019), 'What Country Has the Highest Online Shopping Return Rate?', Retrieved from https://www.paymentsjournal.com/highest-online-shopping-return-rate/

World Commission on Environment and Development: Our Common Future (1987), Retrieved from http://www.un-documents.net/wced-ocf.htm

Designing Closed-Loop Supply Chain Networks with Commercial Returns

2

2.1 INTRODUCTION

Supply chain network design is a key strategic decision in supply chain management as it has a significant impact on supply chain efficiency and profitability. Supply chain engineers and managers work with the existing network to manage inventory and logistics to meet consumers' demand. Supply chain cost, lead time, and risk are some of the common performance measures for supply chain operations. It is mandatory for companies to incorporate aspects affecting the performance of the supply chain operations during the network design stage. A poorly designed or sub-optimal supply chain design will lead to increase in cost and decrease in efficiency of the supply chain operations. Companies are striving to manufacture and distribute the products in an efficient manner to stay highly competitive in today's market. As discussed in Chapter 1, most supply chain networks are receiving huge volumes of returns. The returns trend keeps increasing with increasing online consumer sales, high product variety, and changing consumer preferences. To gain economic advantage by processing these returns and selling refurbished products, companies have to design their supply chains to accommodate both forward and return flows.

Closed-Loop Supply Chains (CLSCs) are the integration of the traditional forward supply chain and the reverse supply chain. Integrating forward and reverse supply chains is a challenging task because of the differences in the entities and the nature of the activities these entities perform that make up the forward and reverse flows. A forward supply chain is a demand-driven flow, whereas a reverse supply chain is a supply-driven flow, in which the supply of return products and its quality is outside the direct control of the company. It would be easier to solve the forward and reverse logistics network design problems independently. However, establishing a reverse network independent of the forward network increases infrastructure costs and reduces the profit potential associated with remanufacturing as it overlooks the interdependence of the forward and reverse flows (Uster *et al.*, 2007). Many researchers have emphasized the importance of addressing forward and reverse supply chains in an integrated manner considering the impact of returns in the network (Fleischmann *et al.*, 2001, Uster *et al.*, 2007, Pishvaee *et al.*, 2010).

This chapter will first introduce and describe a four-stage CLSC network with commercial returns. This network structure will be used for discussion across all the remaining chapters. A modeling framework and an optimization model is presented to design the four-stage CLSC network, incorporating customer acceptance rate for refurbished products, with the objective of maximizing supply chain profit. This model is illustrated and analyzed using a realistic example. A detailed analysis is presented on the effect of return parameters in the supply chain design and its profitability. The analysis is tailored to answer the following two questions: what is the gain in adopting a CLSC with recycling? and how sensitive is your supply chain design and profitability to the return parameters?

2.2 PROBLEM DESCRIPTION AND CONSIDERATIONS

Consider a four-stage CLSC network which produces and distributes a single product and recycles the return products in the return flow path. The set of suppliers is defined by $S = \{1,2...,n_S\}$, the set of manufacturing plants $M = \{1,2... n_M\}$, the set of warehouses $W = \{1,2...,n_W\}$, the set of hybrid facilities $H = \{1,2...,n_H\}$, the set of retailers $C = \{1,2,...,n_C\}$, and the set of recovery centers $R = \{1,2,...,n_R\}$. Let $L = \{1,2,...,n_L\}$ be the set of capacity levels at the warehouses and hybrid facilities. Figure 2.1 shows the CLSC network considered in the study.

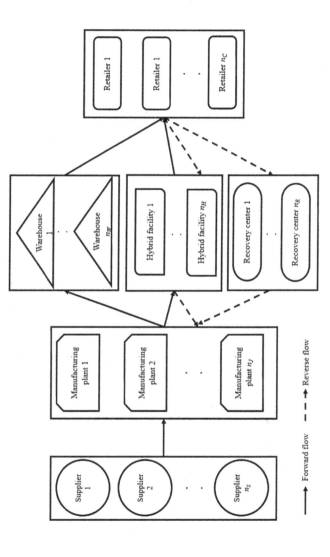

FIGURE 2.1 Structure of the CLSC network.

The CLSC consists of both a forward product flow path and a return product flow path. The forward product flow path comprises suppliers, manufacturing plants, warehouses, hybrid facilities, and retailers. The return product flow path includes retailers, hybrid facilities, recovery centers, and plants. Raw materials that go into the production of new products are supplied by suppliers. We assume that suppliers can produce and supply all the raw materials required for producing a new product. However, we could easily modify the model to consider multiple raw materials and their supply sources. Manufacturing plants can produce new products using the raw materials procured from suppliers and refurbish the products returned by customers. Warehouses are used to distribute both new products and refurbished products in the forward channel. However, warehouses are not equipped with inspection personnel/ equipment to inspect returned products from the retailers in the return channel. Products returned at the retailers should be inspected and tagged for appropriate recovery options. Recovery centers are an entity of the reverse channel or reverse supply chain, which collect return products from the retailers, inspect, and distribute them to the manufacturing plants for refurbishing. Hybrid facilities are a combination of a warehouse and a recovery center. They act as a warehouse in the forward channel and as a recovery center in the reverse channel. Hybrid facilities are more common in practice and offer advantages in terms of significant savings in cost of infrastructure, equipment, and human resources (Easwaran and Uster, 2010). Retailers satisfy the demand in the forward channel and collect the products that are returned by customers. Retailers face a deterministic demand for the product that is satisfied both by using new products and refurbished ones. Customers can choose to buy a new product at a full retail price or a refurbished product at a discounted price.

Customers return used or defective products to retailers. The focus of this book is analyzing CLSCs with commercial returns. These are product returns that can be potentially recovered by *light repair operations or refurbishing*. Customers often have a 90-day period to return a product from the date of purchase for reasons ranging from customer dissatisfaction (due to color, price, etc.) to shipping damage, and defective parts. These returns, named as commercial returns, initiate the reverse flow in the network. Say, for example, a customer buys a laptop (say laptop A) from retailer 1 and after 2 weeks of usage finds a different laptop (say laptop B) with the same specifications as laptop A, for a better price. He decides to return laptop A at retailer 1 and purchase laptop B. This return is categorized as a customer dissatisfaction due to the price of the product. Similarly, other returns in the network due to other reasons such as evaluation product, shipping damage, or defective product are also considered.

The returned products are transported back to the manufacturing plants through the return channel for refurbishing. Products that are not sent for refurbishing could be disposed at the retailer or at the hybrid facilities/recover centers after inspection. The products are refurbished and sent to the retailers through the forward channel. The new products and the refurbished products flowing into the retailers are used to satisfy customer demands. In real-world, not all customers are willing to buy refurbished products (due to its discounted price) fearing inferior product performance during its lifecycle. In this framework, it is assumed that only a proportion of customers are willing to buy the refurbished products.

The network design problem is developed as a single-period, single-product Mixed Integer Linear Programming (MILP) model and is illustrated using a realistic example. The model will be referred to as the *Base Model*. The proposed model is developed to: (a) select the appropriate set of warehouses, hybrid facilities, and recovery center locations in the CLSC network; (b) determine the optimal distribution of new products, return products, and refurbished products through the selected set of facilities; and (c) determine the optimal amount of new and refurbished products produced at the manufacturing plants based on the product return rate and customers' acceptance rate of refurbished products. The objective of the model is to maximize the total profit of the integrated supply chain.

The model in this chapter assumes that there are no categories among the customers (i.e., no high-end and low-end customers). A fraction of the customers (denoted by $\propto$) are willing to buy refurbished products. Also, this chapter does not differentiate the returned products based on its return quality. An average cost is assumed to refurbish the returned items. Demand at the retailers must be satisfied either by new or refurbished products and no shortages are allowed. The refurbished products are assumed to be available for sale within the same planning horizon. The production process could produce defective products, which contributes to the return rate of the products at the retailer. Refurbished products are assumed to be sold/allocated on a first-come, first-served basis, as demand for refurbished products can exceed their supply. The remaining customers continue purchasing new products. The transportation cost of raw material from the supplier to the manufacturing plant is included in the purchasing cost. All the facilities in the supply chain (suppliers, manufacturing plants, warehouses, hybrid facilities, recovery centers) have capacity limitations.

The list of input parameters for the Base Model are discussed as follows: let cap_m be the production capacity at the manufacturing plant m, cap^l_w be the capacity of warehouse w of capacity level l, cap^l_h be the capacity of hybrid facility h of capacity level l, cap_r be the capacity of recovery center r, cap_s be the raw material supply capacity of supplier s, d_c be the demand for products at retailer c, γ be the fraction of demand returned at retailer or customer return rate, and α be the fraction of customers willing to buy refurbished items or customer acceptance rate.

The following are the cost components in the model: Let p_{sm} be the purchasing cost of raw material from supplier s by plant m, tr_{mw} be the transportation cost per unit from plant m to warehouse w, tr_{wc} be the transportation cost per unit from warehouse w to retailer c, tr_{mh} be the transportation cost per unit between plant m and hybrid facility h, tr_{hc} be the transportation cost per unit between hybrid facility h and retailer c, tr_{cr} be the transportation cost per unit from retailer c to recovery center r, tr_{rm} be the transportation cost per unit from recovery center r to plant m, pc_m and rc_m be the new product production cost and return product refurbishing cost at the plants, respectively, np and rp be the selling prices of new and refurbished products, respectively, in_r and in_h be the inspection cost of the returned products at the recovery center and hybrid facility, respectively, f^l_w be the fixed cost of opening a warehouse w of capacity level l, f^l_h be the fixed cost of opening a hybrid facility h of capacity level l, and f_r be the fixed cost of opening a recovery center r.

The following are the decision variables in the model: QSM_{sm} is the quantity of raw materials purchased from supplier s by plant m, QMW_{mw} is the quantity of new products transported from plant m to warehouse w, $RQMW_{mw}$ is the quantity of refurbished products transported from plant m to warehouse w, QMH_{mh} is the quantity of new products transported from plant m to hybrid facility h, $RQMH_{mh}$ is the quantity of refurbished products transported from plant m to hybrid facility h, QWC_{wc} and $RQWC_{wc}$ is quantity of new products and refurbished products transported from warehouse w to retailer c, QHC_{hc} is the quantity of new products transported from hybrid facility h to retailer c, $RQHC_{hc}$ is the quantity of refurbished products transported from hybrid facility h to retailer c, $RQCH_{ch}$ is the quantity of returned products transported from retailer c to hybrid facility h, $RQCR_{cr}$ is the quantity of returned products transported from retailer c to recovery center r, $RQHM_{hm}$ is the quantity of returned products transported from hybrid facility h to plant m, $RQRM_{rm}$ is the quantity of returned products transported from recovery center r to plant m, $DISP_c$ is the quantity of return products disposed at retailer c, δ^l_w is the location selection binary variable for warehouses w with capacity l, η^l_h is the location selection binary variable for hybrid facility h with capacity l, and ξ_r is the location selection for recovery center r.

2.3 THE MIXED INTEGER LINEAR PROGRAMMING MODEL (BASE MODEL)

The framework of the optimization model is as follows: The objective of the model is to maximize the total profit of the CLSC, defined as *total revenue subtracted from the total cost to manufacture/refurbish and distribute products in the supply chain*. Revenue of the supply chain comes from selling new and refurbished products. Total cost of the supply chain includes variable and fixed costs incurred in manufacturing and distributing the product. Variable costs include the cost of purchasing raw materials, cost of manufacturing, transportation cost between the facilities in the supply chain, and inspection cost at the recovery centers and hybrid facilities. Fixed cost includes the cost of opening the warehouses, hybrid facilities, and recovery centers.

The model is subject to the supply capacity restrictions at the suppliers, production capacity restriction at the manufacturing plants, flow conservation constraints for new products at the manufacturing plants, flow conservation constraints for refurbished products constraint at the manufacturing plants, capacity restriction and location selection constraint for the warehouses, flow conservation of new products constraint at the warehouses, flow conservation of refurbished products constraint at the warehouses, capacity restriction and location selection constraint for the hybrid facilities, flow conservation of new products constraint at the hybrid facilities, and flow conservation of refurbished products constraint at the hybrid facilities.

The proposed mathematical model can be accessed from https://www.routledge.com/Design-and-Analysis-of-Closed-Loop-Supply-Chain-Networks/Pazhani/p/book/9780367537494 (see Base Model section). All the equation numbers are references from the web link.

Constraint (2.1) ensures that the raw material supply capacity at supplier s is not violated. The suppliers have a finite supply capacity. Constraint set (2.2) ensures that the sum of products manufactured at the plant m is less than or equal to its capacity. The products produced at plant m are defined by the sum of new and refurbished products transported to warehouses and hybrid facilities from plant m. Constraint set (2.3) represents the flow conservation constraints for new products at manufacturing plant m. The raw material purchased from all suppliers in plant m should be equal to the quantity of new products flowing out of that plant to the warehouses and hybrid facilities. Similarly, constraint set (2.4) is the flow conservation constraints for recycled products at the manufacturer. The quantity of returns flowing into plant m from hybrid facilities and recovery centers should be equal to the quantity of refurbished products shipped out of that plant in the forward chain, to warehouses and hybrid facilities. Constraint set (2.5) represents the capacity and location selection of warehouses. Given the warehouse is selected for operation, this constraint set ensures that the quantity of new products and refurbished products shipped to a warehouse w does not exceed its storage capacity. The company can choose to build the warehouses at any of the capacity levels. Constraint set (2.6) ensures that only one of the capacity levels is selected if warehouse w is opened. Constraint set (2.7) represents the flow conservation constraint at the warehouses. The quantity of new products flowing into warehouse w should be equal to the new products flowing out of that warehouse. Similarly, constraint set (2.8) ensures flow conservation of refurbished products at warehouse w. Hybrid facilities also process product returns from the retailer. Constraint set (2.9) is the capacity constraints at the hybrid facilities. If a hybrid facility h is selected for operation, this constraint ensures that the flow of new and refurbished products and flow of return products into a hybrid facility h does not exceed its capacity. Constraint set (2.10) ensures that only one of the capacity levels is picked, if hybrid facility h is selected. Constraint set (2.11) and (2.12) is the flow conservation constraints for new and refurbished products at hybrid facility h in the forward channel. Constraint set (2.13) is the flow conservation constraint for returned products at the hybrid facility h in the return channel. Constraint set (2.14) represents the capacity and location selection constraints for recovery centers. Constraint set (2.15) is the flow conservation constraints for recovery center r. Constraint set (2.16) is the demand satisfaction constraints. Demand at the retailer c is satisfied with either new or refurbished products. Constraint set (2.17) ensures that total refurbishing activity should be less than or equal to the quantity of refurbished products accepted by customers. Constraint set (2.18) ensures that the returned products at the retailer are either sent to refurbishing, either via recovery centers or hybrid facilities, or are disposed-off at the retailer. Constraint set (2.19) describes non-negativity and binary conditions on the decision variables.

2.4 ANALYSIS OF THE ILLUSTRATIVE EXAMPLE

This section will illustrate the model for the four-stage supply chain using a hypothetical example. The supply chain network has 20 potential suppliers for supplying raw materials to the manufacturing plants. The raw materials go into manufacturing new products. The company has five manufacturing plants for producing new products and refurbishing product returns. The supply chain consists of 16 potential warehousing facilities to distribute new and refurbished products to the retailers in the forward channel. Five potential recovery centers are considered to collect product returns from the retailers, inspect and distribute them to the manufacturing plants. Nine potential hybrid facilities are considered to distribute products in the forward channel and return channel. The network has 100 retailers, who face demand for new and refurbished products from the customers.

All the input parameters for the illustrative example are randomly generated from uniform distributions. Purchasing cost of raw materials from suppliers are uniformly generated from ~unif($600, $800) per unit. Manufacturing and refurbishing costs at the manufacturers are ~unif($25, $35) per unit and ~unif($5, $10) per unit, respectively. Transportation costs between manufacturer and warehouses/hybrid facilities/recovery centers are ~unif($20, $35) per unit. Transportation costs between warehouses/hybrid facilities/recovery centers and retailers are ~unif($45, $55) per unit. Transportation cost increases as the products progress in the supply chain system closer to the retailers. This can be due factors like increase in product value, economies of scale in shipping, etc. Inspection cost at the hybrid facilities/recovery centers for the returned products are ~unif($2, $5) per unit. The demand at the retailers is ~unif(500, 700) units. Both product return percentage and customer acceptance rates are 30% in this illustrative example. New products are sold at the retailers for $1,000 per unit. Refurbished products are sold at 25% discount from the new product price, that is, $750 per unit.

Each potential warehouse and hybrid facility can be opened in one of the three different sizes (small, medium, large). Fixed cost for opening warehouses for the three different sizes are generated from ~unif($350,000, $450,000), ~unif($450,000, $550,000), ~unif($550,000, $650,000). Fixed cost for opening hybrid facilities for the three different sizes are generated from ~unif($450,000, $550,000), ~unif($550,000, $650,000), ~unif($650,000, $750,000). Fixed cost for opening recovery centers ~unif($450,000, $550,000).

Production capacity at the manufacturers are assigned following uniform distribution using ~unif(15,000, 25,000) units as follows: Plant 1 has a capacity of 15,430 units, Plant 2 has a capacity of 23,295 units, Plant 3 has a capacity of 16,381 units, Plant 4 has a capacity of 18,922 units, and Plant 5 has a capacity of 23,583 units. Supplier capacities are uniformly distributed between ~unif(5,000, 10,000) units. Table 2.1 shows the capacities of the suppliers.

Let td be the total demand and tr be the total expected returns across all the retailers. Capacities at the warehouses, for the three sizes, are generated as ~unif(10%, 20%) $\times tot\_dem$, ~unif(20%, 30%) x td, and ~unif(30%, 40%) $\times td$. Capacities at the hybrid facilities, for the three sizes, are generated as ~unif(10%, 20%) $\times (td + tr)$, ~unif(20%, 30%) $\times (td + tr)$, and ~unif(30%, 40%) $\times (td + tr)$. Capacities at the recovery centers are generated

TABLE 2.1   Capacity of the suppliers

	CAPACITY OF SUPPLIERS
Supplier 1	7,109
Supplier 2	6,902
Supplier 3	6,648
Supplier 4	6,578
Supplier 5	7,928
Supplier 6	6,521
Supplier 7	8,631
Supplier 8	6,133
Supplier 9	6,382
Supplier 10	5,359
Supplier 11	7,481
Supplier 12	7,371
Supplier 13	6,857
Supplier 14	9,544
Supplier 15	9,269
Supplier 16	8,800
Supplier 17	7,921
Supplier 18	5,921
Supplier 19	5,079
Supplier 20	9,025

as ~unif(10%, 30%) × *tr*. Table 2.2 shows the capacity and fixed cost of warehouses. Table 2.3 shows the capacity and fixed cost of hybrid facilities. Table 2.4 shows the capacity and fixed cost of recovery centers. Production capacity data and data in Tables 2.1, 2.2, 2.3, and 2.4 will be used throughout the book in the illustrative examples.

The input parameters for the illustrative example are coded in Microsoft Visual C++ 6.0. The mathematical model is coded and solved using a commercial optimization software package. The mathematical model for this illustrative example has 467 constraints and 7,013 variables (with 6,933 continuous and 80 binary variables).

Optimal profit of the supply chain from the model for this example is $21,125,300. We will discuss below the inferences from the mathematical model solution.

Suppliers 1, 5, 6, 8, 9, 10, and 15 are chosen to supply raw materials to the plants for producing new products. Supplier 15 is allocated 22.03% of the total volume, followed by supplier 1 (16.90%), supplier 6 (15.50%), supplier 9 (15.17%), supplier 8 (14.58%), supplier 10 (12.74%), and supplier 5 (3.08%). All the manufacturing plants are used to produce new products. Plants 1, 3,

TABLE 2.2   Capacity and fixed cost of warehouses

	SIZE 1		SIZE 2		SIZE 3	
	CAPACITY	FIXED COST ($)	CAPACITY	FIXED COST ($)	CAPACITY	FIXED COST ($)
Warehouse 1	9,927	415,379	14,108	485,024	21,159	602,485
Warehouse 2	10,189	419,745	15,071	501,072	19,101	568,214
Warehouse 3	9,111	401,781	16,999	533,186	20,208	586,648
Warehouse 4	8,259	387,590	16,147	519,002	21,062	600,878
Warehouse 5	7,338	372,253	16,650	527,372	18,516	558,459
Warehouse 6	6,551	359,148	17,748	545,677	18,610	560,024
Warehouse 7	11,429	440,403	16,342	522,251	18,049	550,683
Warehouse 8	11,905	448,339	17,123	535,253	19,397	573,142
Warehouse 9	8,528	392,081	17,437	540,491	20,912	598,379
Warehouse 10	9,414	406,834	12,146	452,351	18,984	566,264
Warehouse 11	10,872	431,128	16,382	522,916	22,186	619,599
Warehouse 12	6,434	357,184	15,361	505,906	23,119	635,151
Warehouse 13	7,176	369,558	13,283	471,290	23,461	640,838
Warehouse 14	11,765	445,996	14,413	490,108	22,130	618,672
Warehouse 15	7,880	381,276	12,422	456,948	19,374	572,760
Warehouse 16	11,280	437,919	16,099	518,195	23,635	643,737

TABLE 2.3   Capacity and fixed cost of hybrid facilities

	SIZE 1		SIZE 2		SIZE 3	
	CAPACITY	FIXED COST ($)	CAPACITY	FIXED COST ($)	CAPACITY	FIXED COST ($)
Hybrid facility 1	14,431	535,055	19,821	604,168	25,705	679,620
Hybrid facility 2	15,483	548,545	22,142	633,930	24,469	663,762
Hybrid facility 3	15,212	545,063	21,371	624,042	29,541	728,810
Hybrid facility 4	12,053	504,564	21,501	625,711	29,035	722,321
Hybrid facility 5	8,106	453,949	16,957	567,441	29,401	727,003
Hybrid facility 6	10,912	489,933	15,642	550,575	25,261	673,919
Hybrid facility 7	10,877	489,485	21,015	619,481	26,553	690,486
Hybrid facility 8	12,208	506,550	19,956	605,895	26,477	689,520
Hybrid facility 9	11,374	495,855	23,277	648,485	28,849	719,929

TABLE 2.4   Capacity and fixed cost of recovery centers

	CAPACITY	FIXED COST ($)
Recovery center 1	2,879	5,084
Recovery center 2	5,084	541,555
Recovery center 3	3,163	488,090
Recovery center 4	4,943	537,654
Recovery center 5	3,925	509,293

4, and 5 received returned products and are used in refurbishing these return products. There were no warehouses and recovery centers opened for product distribution in the optimal solution. Three hybrid facilities (at locations 6, 7, and 8), each of capacity level 3, are opened. These hybrid facilities distribute new and refurbished products to the retailers in the forward channel. They also collect, inspect, and distribute return products to the manufacturing plants in the return channel. The capacity utilization of these three selected hybrid facilities are greater than 99%. The demands at the retailers are satisfied from the selected hybrid facilities in the forward channel and the returns from the retailers are shipped to these hybrid facilities in the return channel. Hybrid facilities offer economic as well as practical advantages. Supply chain managers prefer having fewer numbers of facilities in the supply chain to reduce operations costs, risks related to monitoring and control of the facility, ease of tracking consignments, reduction in manpower and systems cost, and reduction in logistics and transportation risk in material handling. We can also observe that as the customer return and the acceptance rates are 30% each, there is potential to refurbish and sell all the returned items. As expected, in the solution, the returns were refurbished and sold to the customers.

However, with the objective of maximizing the supply chain profit, the model will recommend refurbishing if and only if it is profitable to refurbish, that is, the model will choose to dispose-off the returned products, if the cost of processing returns and refurbishing outweigh financial benefits from refurbishing. To illustrate this concept, we will now solve the example by setting customer return rate and the acceptance rate to zero. The solution from the models are compared in terms of transportation cost (cost of distributing products in both the forward and return channels), refurbishing cost (sum of inspection cost at the hybrid facilities and recovery centers, and the refurbishing cost at the plants), fixed cost (costs for opening warehouses, hybrid facilities, recovery centers), purchasing cost (raw material purchasing cost from the suppliers), and production cost at the manufacturers. The total profit of the supply chain without recycling was $15,692,200 vs. profit of $21,125,300 with recycling. Analyzing the solutions from the model, we observe that the transportation cost increases in the model with recycling due to inclusion of return and refurbished product flows. Fixed cost increases in the model with recycling, as facilities are opened to process and distribute returned and refurbished products. On the other hand, production cost is lower in the model with recycling as a portion of demand will be satisfied using refurbished products. There is also a huge reduction in raw material usage, that is, reduction in purchasing cost. The cost benefits from production and purchasing costs outweighs the cost of transportation, refurbishing, and additional fixed costs. Managers can use this model to show the benefits of incorporating refurbishing activity in their supply chain and to categorically prove that it is possible to reduce the environmental impact/carbon footprint without compromising on the bottom line, that is, profits.

Note that the cost parameters can vary in practice based on the type of product and industry. In this example, purchasing costs account for 74.21% and transportation costs constitute 15.80% of the total supply chain cost. The quantity of returns and refurbishing activities in the supply chain are determined by customer return rate and acceptance rate parameters, which in turn affect these two major costs (purchasing and transportation) in the supply chain. Purchasing cost decreases with increase in customer return rate and acceptance rate. In our example, 30% of customer demand is satisfied using the refurbished products and raw materials are purchased to produce the remaining 70% of demand. Transportation costs in the supply chain increases with customer return and acceptance rates, as cost is incurred for handling the returns.

2.5 EFFECT OF RETURN PARAMETERS ON SUPPLY CHAIN DESIGN AND PROFITABILITY

Customer acceptance rate and fraction of returns parameters are difficult to estimate in practice compared to other cost parameters. This section discusses the effect of these parameters on the total profit and on the network design and explains why these parameters are important. We will vary customer acceptance rate between 0.1 and 0.50 in increments of 0.01 (41 levels) and fraction of returns between 0.1 and 0.50 in increments of 0.01 (41 levels). Network configuration and all cost and capacity input parameters remain the same as in the illustrative example. A total of $41 \times 41 = 1,681$ cases are generated and executed. Interested readers can refer to Pazhani (2014) for a detailed discussion on this analysis.

2.5.1 Effect of Varying Customer Return Rate and Customer Acceptance Rate on Total Profit

The results from the analysis show that the total profit of the supply chain increases with the increase in the customer return and acceptance rates. Total profit is the highest when customer return and acceptance rates are at its maximum levels. This shows that the total profit increases with increasing refurbishing activity. Total profit is bounded by both customer return rates and acceptance rates. For example, consider the case where customer return rate is equal to 0.2. Supply chain profit increases when acceptance rate increases from 0.1 to 0.2 and remains the same for further increase in acceptance rate greater than 0.2 (i.e., the total profit is bounded by return rate).

Given that the supply chain has economic advantages by adopting refurbishing practices, customer acceptance rate affects the total profit. Both customer return and acceptance rates determine the optimal refurbishing quantities in the supply chain. Thus, designing the supply chain solely based on return rates yields sub-optimal solutions as the customer acceptance rates might be lesser than the return rates. Similar to the illustrative example, monetary benefits from recycling/refurbishing is primarily due to the savings in purchasing cost. Purchasing cost is indirectly proportional to the customer return and acceptance rates values as it decreases the need for purchasing raw materials. Like profit values, the purchasing cost is also bounded by customer return and acceptance rates, as it is indirectly proportional to the refurbishing quantities in the supply chain. Transportation cost increases in the supply chain with increase in customer return and acceptance rate levels due to the increase in distribution of returned and refurbished products.

2.5.2 Variation in Network Design with Varying Customer Return Rate and Acceptance Rate

Supplier selection and selection of optimal sites for distribution are strategic level decisions in the supply chain (Ravindran and Warsing, 2013). This section will discuss the changes in the network design (i.e., supplier selection, locations selected for hybrid and warehousing facilities) under varying $\propto$ and γ levels. There were 52 different scenarios (or combinations) of location decisions for hybrid facilities and warehousing facilities among the 1,681 cases. This clearly shows that the location decision is highly sensitive to the changes in ∝ and γ levels. Let us explain this using Scenarios 3 and 4. For ∝ = 0.14 and γ = 0.14, Scenario 3 is optimal. For Scenario 3, hybrid facility 7 (of size 3), warehouse 12 (of size 3), and warehouse 14 (of size 3) are selected. For ∝ = 0.15 and γ = 0.15, Scenario 4 is optimal. Hybrid facility 3 (of size 3), hybrid facility 7 (of size 3), and warehouse 15 (of size 3) are selected in Scenario 4. A small increase in ∝ and γ values leads to changes in optimal network design. If the company's supply chain design is based on Scenario 3, it will be suboptimal for variations in the ∝ and γ. Consider the case where the company's design is Scenario 3, with ∝ = 0.14 and γ = 0.14. This will lead to loss of potential profit from refurbishing due to insufficient capacity and increase in total supply chain cost due to increase in purchasing, production, and distribution costs. On the other hand, when ∝, γ = 0.13, choosing Scenario 4 results in unutilized capacity in the facilities. Operating in a sub-optimal network design leads to higher costs. In summary, network design is a strategic decision and directly impacts the total profit of the supply chain. ∝, γ has a direct impact on the network design of the CLSC.

2.5.3 Supplier Selection

Supplier selection is one of the most strategic decisions in supply chain management because it provides opportunities to reduce costs and increase profits. It is highly important in industries in which the cost of raw materials and parts account for the highest percentage of the total product cost (Mendoza and Ventura, 2010). According to Weber *et al.* (1991), purchased materials, parts, and services account for up to 80% of the product cost in high technology firms.

We will discuss the supplier selection results for the 1681 cases. Out of the 20 potential suppliers considered, suppliers 1, 3, 5, 6, 8, 9, 10, 15, and 19 were the only suppliers selected in the results. Suppliers 1, 6, 8, 15, and 10 were selected in more than 99.90% of the settings. This indicates that these suppliers are not sensitive to variations in ∝ and γ parameters. Suppliers 9 and 5 were selected in 90% of the settings. This implies that these two suppliers can be used as back up suppliers in situations of higher raw material requirements. Supplier 19 was used in 72.04% and supplier 3 in 39.08% of the settings, implying that selection of this supplier is dependent on ∝ and γ parameters. From the results, it is observed that the number of suppliers in the supply chain decreases as the customer return and acceptance rates increases. This is due to the decrease in raw material requirement. Thus, supplier selection in a CLSC is sensitive to the customer return and acceptance rate parameters.

In summary, customer return rate and acceptance rate parameters are uncertain and difficult to estimate in practice. However, both parameters have a significant effect on the financials of the company. Customer return rates can be predicted, to a certain degree, using the quality level of the product. Customer acceptance rate is based on the customer's perception on the quality of the refurbished product and is highly uncertain and difficult to predict. Given the strategic significance of these parameters in CLSC network design, it is imperative to focus on these parameters.

2.6 A REALISTIC CASE

Companies usually price their products based on the total cost of the product. In this example, we consider the same supply chain network as in Section 2.4. The cost parameters are modeled as a function of the new product cost. Let the average total cost of the product (pp) be $750 and the profit margin be 20%, that is, selling price of the new product is $900. Refurbished products are sold at 25% discount from the new product price. Like the illustrative example, the input parameters are randomly generated from uniform distributions and as a function of new product cost. Purchasing cost of raw materials from suppliers are uniformly generated as ~unif(70%, 80%) of the new product cost (pp). Raw material costs for certain electronic products are almost 70–80% of the total product cost (Ravindran *et al.*, 2010; Ravindran and Warsing, 2013). Manufacturing and refurbishing cost at the manufacturers are ~unif(8%, 12%) × pp and ~unif(14%, 16%) × pp, respectively. Transportation costs between manufacturer and warehouses/hybrid facilities/recovery centers are ~unif(5.5%, 6.5%) × pp. Transportation costs between warehouses/hybrid facilities/recovery centers and retailers are ~unif(8.5%, 9.5%) × pp. Inspection cost at the hybrid facilities/recovery centers for the returned products are ~unif(8.5%, 9.5%) of the average refurbishing cost at the manufacturers.

The demand at the retailers is ~unif(500, 700) units. The capacity of the suppliers, production capacity of the plants, capacities and fixed cost of the warehouses, capacities and fixed cost of the hybrid facilities, and the recovery centers are the same as used in the illustrative example. Product return percentage and customer acceptance rate are set at 30%, initially in this case. They will then be varied, as part of the sensitivity analysis, to study their impact on the final solution.

The data for this case is coded and generated using Microsoft Visual C++ 6.0. The mathematical model is coded and solved using a commercial optimization software package. The optimal profit achieved by the model for this case is $12,217,884. The optimal network design is as follows: Suppliers 1 (16.90%), 3 (3.00%), 4 (0.54%), 5 (7.46%), 6 (15.50%), 7 (20.52%), 8 (14.58%), 9 (15.17%), and 10 (6.33%) are selected to supply raw materials to the plants for producing new products (Note that the percentage of volume allocated to the supplier is shown in the brackets). All the five manufacturing plants are used. Plants 1, 2, 3, and 5 are used in new product production and plants 1, 4, and 5 are used for refurbishing the return products. Warehouse 11 of size 3 is opened. It is used to distribute 15,380 units of new products and 6,806 units of refurbished products. Recovery centers are not used for distribution in the optimal solution. Hybrid facilities 3 and 7, each of size 3, are opened and are utilized to 100% of its capacity. The retailers' demands are satisfied from hybrid facilities 3, 7, and warehouse 11 in the forward channel and the returns at the retailers are shipped to hybrid facilities 3 and 7 in the return channel.

Let's analyze the effect of varying customer acceptance rate, fraction of returns, and refurbishing cost over the total profit and on the network design. Eight levels are considered for customer acceptance rate and customer return rate parameters. They are varied between 0.1 and 0.45 insteps of 0.05. Refurbishing cost at the manufacturer is increased by a factor (rf). rf is varied between 0.01 and 0.89 in increments of 0.02 (45 levels). A total of $8 \times 8 \times 45 = 2,880$ cases are generated and executed. Interested readers can refer to Pazhani (2014) for a detailed discussion on this analysis.

Similar to the results obtained in the earlier section, the results from this case also shows that customer acceptance and return rates affect the total profit and network design. Further, the changes in refurbishing cost also have an effect on total profit of the supply chain. This is intuitive as the total profit decreases with an increase in refurbishing cost for a given refurbishing quantity. Earlier, we established that the refurbishing quantities are bounded customer return and acceptance rates. This analysis shows that the refurbishing cost is also a significant parameter in determining refurbishing quantities, that is, the company might not want to refurbish if the cost of refurbishing outweighs its benefits. What affects the refurbishing cost? It is the quality of the returned products. Can we predict the quality of returned products? Quality of the products is another parameter that is difficult to estimate in practice.

Out of the 2880 cases, we find 62 different combinations of location decisions for hybrid facilities and warehousing facilities. Location selection is also highly sensitive to the changes in $\propto$, γ, and rf levels. As discussed earlier, sub-optimal network design will lead to loss of potential profits. Out of 20 potential suppliers, 1, 3, 4, 5, 6, 7, 8, 9, 10, 15, 17, and 19 appeared in the allocation results of all the 2,880 cases. Suppliers 1, 5, 6, 7, 8, and 9 were selected in all the 2,880 cases. These suppliers are insensitive to the variation in the $\propto$, γ, and *rf* parameters. Other suppliers (3, 4, 10, 15, 17, and 19) were not used in all the cases implying that they are sensitive to $\propto$, γ, and *rf* parameters.

In summary, this case analyzes the effect of refurbishing cost, along with $\propto$ and γ parameters, on total profit and network design decisions. Refurbishing cost is an important cost component that varies significantly based on the quality of the product. It is imperative that we analyze and model the refurbishing cost carefully in the model to better estimate the costs closer to reality.

2.7 MANAGERIAL IMPLICATIONS

The framework proposed in this chapter is generic and can be easily adopted by companies to analyze potential benefits due to recycling in their supply chain network. The model can be used by supply chain managers to design their supply chain network or redesign their existing network incorporating returns. The results from the model on supplier selection will also support the purchasing managers in their price and volume contract negotiations. The model also gives the optimal distribution of the products in the forward and reverse flow paths.

The analysis in this chapter helps us to understand the sensitivity of three important return parameters (return rate, acceptance rate for refurbished products, and refurbishing cost) on CLSC network design. The analysis shows that the customer return rate and acceptance rate for refurbished products have a significant effect on the total supply chain profit. They also affect the optimal configuration of the network, like facility location selection and supplier selection. Further, using a realistic example we show that the changes in refurbishing cost has a significant impact on the total supply chain profit and on the network design. In real world, these return parameters are difficult to estimate. However, their effect on the supply chain design, and profitability is significant.

The Base Model presented in this chapter can be used by supply chain managers for developing an initial network design with forecasted return rate and predicted customer acceptance rate for refurbished products. This can be used as a baseline to propose changes/improvements and to convince upper management. The later chapters in the book will discuss in detail on how to further refine this Base Model to improve profitability and robustness of the supply chain. If companies have a good prediction of their return parameters and the small variations in refurbishing cost, this model can come in handy to design or redesign their supply chain.

2.8 SUMMARY

In this chapter, we address these questions: (a) what is the gain in adopting a CLSC with recycling? and (b) how sensitive is your supply chain design and profit to the return parameters?. We study a four-stage CLSC network and propose solutions for the optimal design of the network. A formal integrated optimization model of a supply chain, where materials are recycled in the return flow path, is developed as a mixed integer linear programming model. This model is called the *Base Model*. The MILP model is used to determine the optimal location of facilities, supplier selection, and distribution of new, refurbished and return products in the CLSC network.

In the *Base Model*, we assume that there are no categories among the customers and return items. The returned products are assumed to be of similar condition and thus an average refurbishing cost is considered, irrespective of the condition of the product. Firstly, an illustrative example is presented using hypothetical data and solved using the model. The optimal solution from the model is discussed. Initially, we perform sensitivity analysis to study the effect of varying customer return rate and acceptance rate on the total profit of the supply chain and the network design. The results show that these return parameters have a significant effect on the total profit. Also, the changes in return parameters lead to changes in supplier selection decisions, location selection decisions of warehouses, hybrid facilities, and recovery centers. Next, we also present a realistic case, where the cost parameters are modeled as a function of the new product cost. Using this case, we perform sensitivity analysis to study the effect of refurbishing cost on total profit and network design decisions. The results from the analysis show that the refurbishing cost has a significant impact on total profit and network design decisions.

All the three return parameters, refurbishing cost, customer return rate, and acceptance rate for refurbished products, are difficult to predict in the real world. However, it has a strategic significance in CLSC network design. Chapter 4 proposes and analyzes an *Extended Base Model*, which improves the modeling phenomenon for refurbishing cost and customer acceptance rates.

2.9 MULTI-OBJECTIVE EXTENSIONS

Real-world problems are often multi-objective in nature. The objectives can differ from industry to industry. For example, a company can have objectives to maximize profit and responsiveness. Another company could focus on reducing disruption risk, profit, and fulfilment rate. The model presented in this chapter is generic and can be analyzed by including multiple objectives such as supply chain risk, carbon footprint, energy usage, and responsiveness in forward and reverse supply chains. A multi-objective extension of the Base Model can be read from Pazhani and Ravindran (2018). This paper focuses on designing an optimal CLSC network with the objectives of maximizing the supply chain profit and minimizing the energy usage at the warehousing facilities and energy consumed during transportation.

REFERENCES

Easwaran, G., Uster, H. (2010), 'A closed-loop supply chain network design problem with integrated forward and reverse channel decisions', *IIE Transactions* **42**(11), 779–792.

Fleischmann, M., Beullens, P., Bloemhof-Ruwaardz, J. M., Wassenhove, L. N. (2001), 'The impact of product recovery on logistics network design', *Production and Operations Management* **10**(2), 156–173.

Mendoza, A., Ventura, J. A. (2010), 'A serial inventory system with supplier selection and order quantity allocation', *European Journal of Operational Research* **207**(3), 1304–1315.

Pazhani (2014). Design of Closed-Loop Supply Chain Networks, The Pennsylvania State University, https://etda.libraries.psu.edu/files/final_submissions/9801

Pazhani, S., Ravindran, A. R. (2018), 'A bi-criteria model for closed-loop supply chain network design', *International Journal of Operational Research* **31**(3), 330–356.

Pishvaee, M. S., Farahani, R. Z., Dullaert, W. (2010), 'A memetic algorithm for bi-objective integrated forward/reverse logistics network design', *Computers & Operations Research* **37**(6), 1100–1112.

Ravindran, A. R. Warsing, D. P. (2013), *Supply Chain Engineering: Models and Applications*, CRC Press, Boca Raton, FL.

Ravindran, A. R., Bilsel, U. R., Wadhwa, V., Yang, T. (2010), 'Risk adjusted multi-criteria supplier selection models with applications', *International Journal of Production Research* **48**(2), 405–424.

Uster, H., Easwaran, G., Akçali, E., Cetinkaya, S. (2007), 'Benders decomposition with alternative multiple cuts for a multi-product closed-loop supply chain network design model', *Naval Research Logistics* **54**(8), 890–907.

Weber, C. A., Current, J. R., Benton, W. C. (1991), 'Vendor selection criteria and methods', *European Journal of Operational Research* **50**(1), 2–18.

Characteristics of Commercial Return Products

3

3.1 INTRODUCTION

In the Base Model presented in Chapter 2, we assumed that the returns in the network are similar in quality and it takes the same cost/effort to refurbish these return products at the manufacturer. We also assumed that a fraction of customers would be willing to buy refurbished products if refurbished products are available in the market. These customers are willing to buy refurbished products due to the price discount offered for these products. In Chapter 2, we showed that these parameters have a direct impact on profitability and network design of the supply chain. In this chapter, we will analyze these assumptions and propose realistic methods for modeling refurbishing cost and customer willingness to buy refurbished products.

Firstly, let's analyze the assumption of constant refurbishing cost across all returned products, which is not always true in practice. The products returned at the retailer are inspected and sent to a refurbishing facility for refurbishing. The company can choose to refurbish returned products internally (at their manufacturing facility) or use a third-party facility. In either case, refurbishing cost is directly proportional to the defects in the product returns. The quality of product returns varies depending upon the length of product usage, type of defect (functional vs cosmetic), defect/damage intensity, damages due to shipping, etc., for example, replacing a broken key in a laptop keyboard vs. replacing the cracked screen. This categorization is important so as (1) to plan resources for refurbishing, such as labor, transportation, storage space, parts, machine time, etc., and (2) to identify product quality issues and improve them in the products.

Secondly, let's consider the customer acceptance rate for refurbished products. In practice, customer acceptance of refurbished products does not always depend on the discount offered. Even the price-sensitive customers are willing to buy refurbished products only if their utility of that product is positive, that is, their perceived cost for the product is higher than the discount offered for the product. There are also a set of customers who are not willing to buy refurbished products due to ambiguity in product quality, both cosmetically and in functionality. Consider a critical electronic product in an automobile. There could be no customers willing to buy refurbished product due to its criticality in use and ambiguity in the refurbished product quality. Customer acceptance behavior is also an important factor in network design as it directly affects the network design and profitability of the supply chain.

This chapter is organized as follows: Section 3.2 presents a discussion on refurbishing cost as a function of product quality. It also discusses a refurbishing cost categorization from literature. Section 3.3 gives a brief discussion on the customer categories and their perception toward refurbished products. This section also introduces a customer categorization based on the consumers' buying behavior toward buying refurbished products – low-end customers and high-end customers. Sections 3.4 and 3.5 discusses modeling the acceptance rates for low-end and high-end customers. Hypothetical datasets are used to illustrate these models. Section 3.6 presents the conclusions of this chapter.

3.2 MODELING REFURBISHING COST

In the Base Model presented in Chapter 2, we assumed that the returns in the network are similar in quality and it takes same cost to refurbish these return products at the manufacturer. This assumption is not always true in practice and in most of the instances the quality of product returns varies depending on the length of product usage, type of defect (functional vs cosmetic), defect/damage intensity, damages due to shipping, etc. Some studies in the literature have addressed this concept of modeling refurbishing cost as a function of return product quality. A study by Ovchinnikov (2011) categorizes return products into four tiers based on the return product condition (quality) and the associated refurbishing cost: (1) Tier 1 returns are product returns that did not meet the expectation of the customers, for a range of functional, cosmetic, pricing, or even feel reasons. For instance, the customers could have found the same product for a lower price, or the customer might have disliked the color of the product. (2) Tier 2 returns have some shipping damages or minor product defects. (3) Major product defects are categorized as Tiers 3 and 4, based on the extent of defect/damage. Tier 4 defects/damages are considered to consume more resources and time to refurbish than Tier 3 (i.e., the cost of refurbishing Tier 4 returns is higher than the Tier 3 returns). For example, consider a defective mobile phone. Tier 3 damage can be a malfunctioning camera as opposed to Tier 4, which can be a cracked screen. However, this four-tier categorization does not apply across all products/industries. It depends on the product, defects in return products, and the associated cost for refurbishing the returned product. For example, consider a laptop and a computer monitor – a laptop could potentially encounter defects in various parts requiring different refurbishing costs than that of a computer monitor. Data on returns at the retailers and the associated refurbishing cost (or potential repair cost if the company has not started refurbishing activities) could be used to model these categories. Refurbishing includes *inspection cost*, cost of inspecting the returned product (to identify defects and plan for appropriate refurbishing), and *refurbishing cost*: the cost of performing the actual refurbishing.

We will now give an overview of the four tiers of returns. Open-box products with small cosmetic repairs, but with full functionality are categorized as Tier 1 returns. Labor cost for functional testing and visual inspection for cosmetic defects along with collection and storage cost make up the inspection costs for Tier 1 returns. Tier 1 returns incur the least refurbishing cost. Shipping damages, major cosmetic blemishes, and minor product defects are categorized as Tier 2. Inspection cost includes labor cost for inspecting these shipping damages, electrical and mechanical faults, visual inspection for cosmetic defects, collection, and storage cost. Tier 2 returns typically require damaged-part replacement (material and labor), cleaning and cosmetic repairs, repackaging and restocking cost, and cost of final quality inspection. Major product defects are categorized under Tiers 3 and 4. Even though the proportion of these returns is small, the cost incurred in refurbishing them is high. These returns often need replacing defective parts (material and labor), cost of cleaning and cosmetic makeover, repackaging, and the cost of final quality inspection. Return product inspection for Tiers 3 and 4 returns is comprehensive compared to the other return tiers, as more focus must be given to identify the defects and to determine refurbishing requirements. Refurbishing costs for Tiers 3 and 4 returns are the highest.

A report by Accenture in (2011) finds that Tier 1 returns account for 95% total returns. Only 5% of the returns (in Tiers 2, 3, and 4) is due to actual defects in the product. The case study of a major wireless carrier in North America (Ovchinnikov, 2011) finds Tier 1 returns to be 60% of the total returns. However, both Tier 1 and Tier 2 returns accounted for higher than 90% of the total returns. The difference in opinion across these studies also reinforces our claim that the categorization of returns varies based on the product/industry. Data analysis should be performed to determine return categories for the product/industry under consideration.

How do we incorporate this in the context of supply chain design? The *Base Model* will be modified to consider returns under different tiers, and refurbishing costs for these tiers of returns. The extended model in Chapter 4 will discuss this in detail.

3.3 CUSTOMER CATEGORIES AND THEIR PERCEPTION TOWARD REFURBISHED PRODUCTS

Refurbished products are intended to be fully functional and cosmetically at par with new products. In an ideal case, there is no difference between new and refurbished products. But, what affects perception of consumers toward buying refurbished products? Seller reputation: entity refurbishing the product (original equipment manufacturer [OEM] and their authorized facilities or third-party refurbishing plants) has a positive impact on consumers' perception toward buying refurbished products. Subramanian and Subramanyam (2012) finds that seller's reputation is an important factor in buying remanufactured products. The authors also find that the consumers are even willing to pay a relatively higher price for products remanufactured by OEMs or their authorized facilities. Apart from this, other studies (Guide and Li, 2010; Hazen *et al.*, 2012; Ovchinnikov, 2011; Van Weelden *et al.*, 2016) have shown two other factors contributing to consumers' perception toward refurbished products: transparency in refurbishing processes and refurbished product quality or risk of product performance perception.

An empirical study by Hazen *et al.* (2012) shows that the refurbishing processes are a source of ambiguity for consumers. This study finds evidence that consumer's tolerance for the refurbishing process ambiguity and their willingness to pay for refurbished products are directly related. This also affects the consumer's perceived quality of the refurbished product. Mugge *et al.* (2017)

find that the consumer's perception toward buying refurbished products can be improved by providing information about the refurbishing processes. The findings from Subramanian and Subramanyam (2012) also suggest that refurbishing by OEM or their authorized facilities reduces ambiguity for consumers, leading to increased willingness to pay for refurbished products. Amazon's renewed product listing adds the following details in an image as the product journey: (1) customer bought a new product; (2) returned the product with minimal or no signs of wear and tear; (3) checked and restored to original working condition by professionals; and (4) sold as certified refurbished product. This information reduces the ambiguity for consumers on refurbishing products and improves customer perception toward buying refurbished products.

Is price always a factor for buying refurbished products? The answer is: it depends on the type of product/industry, and customer behavior. A survey by Xinhua News Agency (Hazen et al. 2012) finds that 78.1% of the respondents were not willing to buy remanufactured automobile parts. These parts were 50% cheaper than that of the new products. Another research study (Mugge et al., 2017), based on an online survey, finds that only 46% of the sample showed a positive attitude to buy refurbished smartphones. This study also finds that perceived risk of performance and personal innovativeness had a negative impact toward buying refurbished products. A relatively small percentage of consumers were willing to buy refurbished products owing to environmental consciousness. One of the major bottlenecks in the development of the refurbishing industry is the low purchase intentions of consumers for refurbished products (Mugge et al., 2017).

Ovchinnikov's (2011) study classifies customers into high-end and low-end. His study states that the consumers use the price of the refurbished product to judge its quality. There is a common belief that a low-end customer or price-sensitive customer is willing to buy refurbished items owing to its lower price. Ovchinnikov (2011) study assumes that the low-end customers are more inclined toward buying refurbished items (if it is available in the market) due to its lower price. However, Guide and Li (2010) had shown empirically that the consumers' preconceived impressions on the product quality, performance, and durability significantly affects the customer acceptance rate for refurbished products. The consumers will also stay away from refurbished products if he/she had a negative experience with refurbished products in the past. In summary, the objective of a price-sensitive (low-end) customer is to maximize his own utility based on his perceived value on the quality of the refurbished product. This concept is termed as *risk aversion*, where customers face critical uncertainties and ambiguities concerning product quality (Lee, 2011). Risk aversion is defined as a method to avoid potential future regrets in a natural decision-making process, in a system where decision makers perceive uncertainties and ambiguities. The perceived value with respect to its product quality is characterized using a concave function by Okada (2010).

High-end or quality-sensitive consumers are more often inclined toward purchasing new products, owing to quality concerns of refurbished products. Ovchinnikov (2011) finds that the discount offered on refurbished products will attract only a fraction of high-end consumers. The fraction of consumers who switch from new to refurbished products follows an inverted U-shaped curve. At low discount levels, the high-end customers are inclined toward new products as they are not willing to trade-off quality for the small gain in price. A small fraction of consumers, not all, are willing to buy refurbished products, at moderate discount levels, because of savings in price. At large discount levels, the high-end customers preconceive a low product quality and will be more inclined toward buying new products. The product market in which the company operates in the Ovchinnikov (2011) study has high-end and low-end customers. The discounts offered for the refurbished products affect the acceptance rate of the refurbished products. The acceptance rate in the network has to be computed depending on the fraction of these customer categories and their behavior. The customer categories are not generic and can differ across industries and products. Companies and network design managers designing their supply chain should determine the data for their product market/industry. Mugge *et al.* (2017) stress the importance of the companies to have knowledge on the different customer groups and their attitudes toward refurbished smartphones. Subramanian and Subramanyam (2012) find that OEMs are reluctant toward refurbishing products due to the lack of knowledge of consumers' perceptions toward buying refurbished products and their willingness to pay. Given the economic and environmental benefits of refurbishing, determining customer categories and their behavior is an important aspect of Closed-Loop Supply Chain (CLSC) adoption.

Our analysis in Chapter 2 showed that the customer acceptance rate has a significant effect on supply chain profits and network design. Chapter 4 will incorporate the customer categories based on the Ovchinnikov (2011) study in the network design model.

3.4 MODELING LOW-END CUSTOMER ACCEPTANCE RATE

This section will discuss a method to model low-end customer acceptance rates. Let rp be the discounted price of the refurbished products, q be the quality of the product, and $v(q)$ be the perceived value of the product by the consumers for the quality level q. As discussed earlier, a low-end customer tends to buy refurbished products only if the price of the refurbished product is less than the perceived value of the product (i.e., their utility is positive). We consider a scale of 0 and 1 for both q and $v(q)$. When $q = 0$ (i.e., when customers think that the refurbished product has no functional use), then $v(q) = 0$. When $q = 1$ (i.e., when customers perceive that the quality of a refurbished product and a new product is the same), then $v(q) = 1$. The quality level of the product and the perceived value of the product for that quality level is modeled as a concave function.

Let the price of the new product be np. $v(q).np$ gives the customer's perceived value for the refurbished product. Without loss of generality, we consider that $q = 1$ will not exist in the market and customer's perceived value for refurbished products will be less than the new product price, that is, $v(q).np < np$. A low-end customer intends/will buy a refurbished product if the discounted refurbished product price is less than his perceived price (i.e., $v(q).np - rp > 0$).

It is unrealistic to assume a specific value for product quality in the market. We will assume that the customer's perception on product quality is uncertain and has a range $[q - i, q + i]$, with a corresponding perceived value with range $[v(q - i), v(q + i)]$. Based on studies (Atasu *et al.*, 2008; Ferguson, 2009; Mitra, 2007), the customer's willingness-to-pay or the distribution of the perceived value $[v(q - i), v(q + i)]$ is modeled as a uniform distribution. Let $v(x)$ be the perceived value function, which is uniformly distributed between $[v(q - i), v(q + i)]$. Customer acceptance rate for refurbished products is equal to the percentage of customers whose perceived value is greater than rp and can be computed as the probability that of $v(x).np > rp$ and $v(x)$ is uniformly distributed between $[v(q - i), v(q + i)]$. The probability that $v(x).np > rp$ is given by the equation below:

$$P\big(v(x).np> rp\big)=\left(\frac{v(q+i)-\dfrac{rp}{np}}{v(q+i)-v(q-i)}\right) \tag{3.1}$$

When the ratio of refurbished product price to new products price is set at the maximum perceived value, $v(q + i)$, the customer acceptance rate is 0. When the ratio of refurbished product price to new products price is set at the minimum perceived value, $v(q - i)$, the customer acceptance rate is 1. The perceived value of the refurbished product, $v(q)$, is a monotonically increasing curve with q, that is, with increase in uncertainty in refurbished product quality perception $[q - i, q + i]$, there is also a corresponding increase in uncertainty of the perceived value of the product. We can also observe from Equation (3.1) that, for a given refurbished product price, the customer acceptance rate decreases with an increase in uncertainty in refurbished product quality perception. Let's consider a hypothetical data set: {Product quality (q) and perceived value ($v(q)$)} pairs are as follows: {0.00, 0.00}, {0.10, 0.32}, {0.20, 0.45}, {0.30, 0.55}, {0.40, 0.63}, {0.50, 0.71}, {0.60, 0.77}, {0.70, 0.84}, {0.80, 0.89}, {0.90, 0.95}, {1.00, 1.00}. The customer acceptance rate is illustrated using the hypothetical data. Let the refurbished product quality perception in the market be [$q - i = 0.30$, $q + i = 0.7$]. From the hypothetical data set, the perceived value of the product is between [$v(q - i) = 0.55$, $v(q + i) = 0.84$]. The company offers a 25% discount on refurbished products. For this example, the new product is priced (np) at \$1,000 and the refurbished product is priced (rp) at \$750. Substituting the values in Equation (3.1), we get the low-end customer acceptance rate of 31.03% at a 25% discount. We also plot and analyze the customer acceptance rate against different levels of refurbished products discounts. Firstly, the customer acceptance rate linearly increases with an increase in discount offered for the refurbished products. For the hypothetical data set, the customer acceptance rate is 0% until a discount rate of 16%. It linearly increases (greater than 0%) with increase in discount rates greater than 16%. The customer acceptance rate reaches 100% at a discount rate of 45%. It remains at 100% for any discount rates greater than 45%.

Equation (3.1) and the hypothetical data used in this section will be used in Chapters 4 and 5 for modeling low-end customer acceptance rates.

3.5 MODELING HIGH-END CUSTOMER ACCEPTANCE RATE

This section will discuss a methodology to model high-end customer acceptance rates. High-end or quality-sensitive customers value product quality more than the price. Ovchinnikov (2011) finds that the high-end customers use price as a factor to determine the quality level of the product. As discussed in Section 3.3, only a fraction of high-end customers will intend to buy refurbished products at moderate discount levels. This buying behavior of the high-end customers, as a function of the refurbished product price, is modeled as an inverted U-shaped curve. Let θ denote the discount offered for refurbished products, which is equal to $(1 - (rp/np))$. The customer acceptance rate is 0% between 0% to x%, as the customers are not willing to compromise quality till $x\%$ monetary gain. A portion of customers are willing to buy refurbished products for discount offered between $x\%$ to $(x + a)\%$. However, the customer acceptance rate goes back to 0% for any discounts offered between $(x + a)\%$ and 100%. This is because the high-end customers suspect poor quality at this discount level. Let's consider a hypothetical example: {% discount offered for refurbished products and % of high-end customers accepting refurbished products} pairs are as follows: {0%, 0%}, {10%, 0%}, {15%, 0%}, {20%, 6%}, {23%, 12%}, {25%, 15%}, {40%, 19%}, {45%, 16%}, {50%, 14%}, {55%, 12%}, {60%,10%}, {65%,5%}, {70%, 2%}, {75%, 0}. This discount offered vs. customer acceptance rate data follows an inverted U-shaped curve. One way of collecting this data for a product/market is through customer surveys. These surveys are normally done using a discrete set of discounts and the customers' willingness to buy at that discount level. Data from the survey is then summarized to understand consumers buying behavior. We fit a polynomial function over this data to study the customer behavior over a continuous scale of discount values. Equation 3.2 shows the polynomial fit equation for the hypothetical data:

Customer acceptance rate (high-end customers)

$$= 0.008 - (1.227 \times \theta) + \left(13.45 \times \theta^2\right) - \left(31.12 \times \theta^3\right) + \left(20.68 \times \theta^4\right) \tag{3.2}$$

This function can be used to calculate high-end customer acceptance rates for a given refurbished product discount. Consider that the company is offering a 28% discount rate for refurbished product. The acceptance rate for high-end customers (based on the hypothetical data), calculated using Equation (3.2), is 16.29%. Equation (3.2) and the hypothetical data used in this section will be used in Chapters 4 and 5 for modeling high-end customer acceptance rates. Interested readers can refer to Pazhani (2014) for a detailed discussion on modeling low-end and high-end customer acceptance rates.

3.6 SUMMARY

The sensitivity analysis presented using the *Base Model* (in Chapter 2) clearly shows that refurbishing cost and customer acceptance rate parameters have a significant effect on the profitability and the network design of the supply chain. In this chapter, we focus on realistically modeling these two key characteristics. Often these parameters are aggregated in network design models, like we show in the *Base Model*, due to the lack of understanding of financial implications in operating with a sub-optimal network design. Firstly, we briefly discussed the relationship between the condition of the returned products and the cost of refurbishing. We presented a four-tier categorization of product returns, based on the nature and intensity of defect, and the corresponding refurbishing costs. The return categories can vary based on several factors like type of product, industry, market in which they operate, etc. However, this categorization and data is critical for companies to effectively manage their refurbishing operations (i.e., resource planning, parts management, etc.). This categorization is also important for the network design teams to optimally design/operate their supply chains and manage their refurbishing strategy. For example, the company can choose to sell certain return categories to an external vendor or dispose-off at the point of return, which is not profitable to refurbish in the companies' network.

We then discussed the categories of customers and their buying behavior toward buying refurbished products. Customer perception toward buying refurbished products depends on seller reputation, refurbishing process ambiguity, and product discount. We present a scenario with two customer categories – low-end and high-end. Buying behavior of these customer categories vary significantly. Low-end customers are price-sensitive and are willing to purchase the refurbished product if their own utility is positive (i.e., the price of the refurbished product offered is less than their perceived price for that product). On the other hand, high-end customers are quality-sensitive and are not willing to trade quality for financial gains. We present a methodology to model the buying behavior of these two customer categories. Low-end customers are modeled using perceived quality vs. perceived value function. The perceived value function is also modeled as a uniform distribution, to incorporate uncertainty in the market. Given the distribution of the perceived value for the refurbished product in the market and the refurbished product discount, the customer acceptance rate is computed as the probability of customers whose perceived value greater than the refurbished product price. Buying behavior of high-end customers is modeled as a function of discount offered for refurbished products, using an inverted U-shaped curve as follows: the high-end customers are not willing to buy refurbished products at low-discount rates, as they do not want to sacrifice quality for a small monetary gain. A fraction of customers is willing to buy refurbished products at moderate discount levels. At high-discount levels, customers suspect poor quality and are not willing to buy refurbished products. We use hypothetical data to illustrate the customer acceptance rate computations. Customer categories can also vary depending on industry/product and market in which they operate. Studies have shown that understanding customer acceptance behavior is critical for the success of refurbishing activities.

In Chapter 4, we will extend and analyze the *Base Model* by incorporating the product-quality-based refurbishing cost, customer categories, and customer buying behavior in the network design model.

REFERENCES

Atasu, A., Sarvary, M., Wassenhove, L. N. (2008), 'Remanufacturing as a marketing strategy', *Management Science* **54**(9), 1731–1746.

Ferguson, M. E. (2009), 'Strategic and tactical aspects of closed-loop supply chains, foundations and trends in technology', *Information and Operations Management* **3**(2), 103–200.

Guide, V. D. R. Jr., Li, J. (2010), 'The potential for cannibalization of new products sales by remanufactured products', *Decision Sciences* **41**(3), 547–572.

Hazen, B. T., Overstreet, R. E., Jones-Farmer, L. A., Field, H. S. (2012), 'The role of ambiguity tolerance in consumer perception of remanufactured products', *International Journal of Production Economics* **135**(2), 781–790.

Lee, C. (2011), 'Analysis of Decision-Making in Closed-Loop Supply Chains', PhD thesis, Georgia Institute of Technology, Atlanta, GA.

Mitra, S. (2007), 'Revenue management for remanufactured products', *Omega* **35**(5), 553–562.

Mugge, R., Jockin, B., Bocken, N. (2017). How to sell refurbished smartphones? An investigation of different customer groups and appropriate incentives. *Journal of Cleaner Production*, **147**, 284–296.

Okada, E. M. (2010), 'Uncertainty, risk aversion, and wta vs. Wtp', *Marketing Science* **29**(1), 75–84.

Ovchinnikov, A. (2011), 'Revenue and cost management for remanufactured products', *Production and Operations Management* **20**(6), 824–840.

Pazhani (2014). Design of Closed-Loop Supply Chain Networks, The Pennsylvania State University, https://etda.libraries.psu.edu/files/final_submissions/9801

Subramanian, R., Subramanyam, R. (2012), 'Key factors in the market for remanufactured products', *Manufacturing & Service Operations Management* **14**(2), 315–326.

Van Weelden, E., Mugge, R., Bakker, C. (2016). 'Paving the way towards circular consumption: Exploring consumer acceptance of refurbished mobile phones in the Dutch market', *Journal of Cleaner Production*, **113**, 743–754.

Network Design Incorporating Quality of Returns and Customer Acceptance Behavior

4

4.1 INTRODUCTION

The teams designing supply chains within companies should ensure that their design incorporates all relevant parameters that directly affect the operational performance indicators of the supply chain. For instance, supply chain profitability and responsiveness are typical operational performance indicators, which are reviewed continuously and are significantly dependent on the design of the supply chain. Parameters such as supplier lead times and their on-time delivery performance, distribution modes, and their delivery speed have to be considered while building a responsive supply chain. In the context of Closed-Loop Supply Chain (CLSC), we show in Chapter 2, using the *Base Model*, that refurbishing cost and customer acceptance rate have a significant impact on the profitability and network design of the supply chain. However, the *Base Model* assumes one category of returns and the average cost associated with refurbishing these returned products. It also assumes that a fraction of customers is willing to buy refurbished products. Chapter 3 discussed in detail about the classification of the product returns in the supply chain based on their quality, and the customer perception toward buying refurbished products. This chapter will extend the *Base Model* by incorporating return quality-based refurbishing cost and customer acceptance behavior in the network design model. In the modeling framework, we will use four tiers of return categories (Tiers 1, 2, 3, 4) and two categories of customers (high-end and low-end), as defined by Ovchinnikov (2011). However, we can easily modify the model to consider *n* categories of returns and *m* customer categories. For instance, companies can categorize customer behavior into high-end, nominal, and low-end customers. The buying behavior of the customers toward refurbished products can differ by retailers or a group of retailers. These modifications can be easily incorporated within the presented framework.

The remainder of this chapter is organized as follows: Section 4.2 presents the Extended Base Model using mixed integer linear programming. In Section 4.3, we present an illustrative example to show the application of the Extended Base Model. In Section 4.4, sensitivity analysis is carried out to study the effect of fraction of low- and high-end customers, customer return rates, and discounts offered for refurbished products on the total profit of the supply chain. Section 4.5 discusses the managerial implications of the Extended Base Model. Section 4.6 offers the conclusions of the chapter. Multi-objective extensions are discussed in Section 4.7.

4.2 THE MIXED INTEGER LINEAR PROGRAMMING MODEL (EXTENDED BASE MODEL)

This section presents the mixed integer linear programming (MILP) model for the Extended Base Model. The problem description is the same as discussed in Chapter 2. The model is extended by considering four tiers of returns (Tiers 1, 2, 3, 4) and the two categories of customers (high-end, low-end) and their buying behavior, as discussed in Chapter 3. We will now present the additional input parameters, cost components, and decision variables specific to the Extended Base Model, apart from the ones discussed in Chapter 2. Let θ be the discount rate offered for refurbished products, q be the product quality and $v(q)$ be the perceived value for quality level q, $[q - i, q + i]$ be the customer's perception range on product quality in the market, $[v(q - i), v(q + i)]$ be the perceived value of the product for customer's perception range $[q - i, q + i]$, γ be the fraction of demand returned at retailer, β_k be the fraction of Tier k returns.

Customer acceptance rate or the fraction of customers willing to buy refurbished items is defined for each customer category. Let α_1 and α_2 be the fraction of low- and high-end customers willing to buy refurbished items, ω_1 and ω_2 be the fraction of low- and high-end customers. Refurbishing and inspection costs are modified to include tiers of returns (which are classified by the quality of returns). Let rc_{km} be the refurbishing cost for a returned product in Tier k at plant m, in_{kr} and in_{kh} be the inspection cost of returned product in Tier k at recovery center r and hybrid facility h, respectively.

The decision variables related to returns in the return flow are redefined as follows:

$RQCH_{kch}$ is the quantity of returned products in Tier k transported from retailer c to hybrid facility h; $RQCR_{kcr}$ is the quantity of returned products in Tier k transported from retailer c to recovery center r; $RQHM_{khm}$ is the quantity of returned products in Tier k transported from hybrid facility h to plant m; $RQRM_{krm}$ is the quantity of returned products in Tier k transported from recovery center r to plant m; $DISP_{kc}$ is the quantity of return products in Tier k disposed at retailer c.

The framework of the optimization model is as follows: the objective of the model is to maximize the total profit of the CLSC, defined as *total revenue subtracted from the total cost to manufacture/refurbish and distribute products in the supply chain.*

The proposed Extended Base Model can be accessed from https://www.routledge.com/Design-and-Analysis-of-Closed-Loop-Supply-Chain-Networks/Pazhani/p/book/9780367537494 (see Extended Base Model section). All the equation numbers are references from the web link.

Constraint set (4.1) ensures that the quantity of returned products under all tiers flowing into plant m is equal to the quantity of refurbished products flowing out of that plant to the warehouses and hybrid facilities. Constraint set (4.2) ensures that the quantity of new products and refurbished products flowing in the forward channel and returned products flowing in the return channel into a hybrid facility h does not exceed its storage capacity, if the hybrid facility is selected for operation. Constraint set (4.3) ensures that the return product quantity, under all tiers, flowing into the hybrid facility h from the retailers is equal to the amount flowing out of that hybrid facility to the manufacturing plants. Constraint set (4.4) is the location selection and capacity constraint for recovery center r. This constraint ensures that the return products, under all tiers, flowing into the recovery center r from the retailers, are less than or equal to the storage capacity of recovery center r. Constraint set (4.5) is the flow balance constraint for recovery center r. Constraint set (4.6) ensures that the total refurbished products flowing into retailer c are less than or equal to the sum of customer acceptance rate of low-end and high-end customers to buy refurbished products. Constraint set (4.7) is the flow balance constraints at the retailers for the return products in each tier. These return products are disposed-off at the retailer or sent for refurbishing to the manufacturing plants through hybrid facilities and recovery centers. Constraint set (4.8) describes non-negativity conditions for the decision variables.

4.3 ILLUSTRATIVE EXAMPLE – EXTENDED BASE MODEL

We consider the same four-stage supply chain network as discussed in Chapter 2, Section 2.2. The cost parameters are modeled as a function of the new product price similar to the realistic case discussed in Chapter 2 (see Section 2.6). The capacities of the suppliers and plants, capacities and fixed cost of the warehouses, hybrid facilities, and the recovery centers are also the same as given in Chapter 2 (see Section 2.4 and Tables 2.1, 2.2, 2.3, and 2.4).

Refurbishing and inspection costs of returned products are set based on their tier of return, that is, higher the tier of return, higher is the cost of refurbishing and inspection. Initially, we assume a mix of 70% low-end and 30% high-end customers. This will be varied under sensitivity analysis in Section 4.4. Let pp be the cost of a new product. Refurbishing cost at the manufacturers for Tiers 1, 2, 3, and 4 are ~unif(6%, 8%)*pp, ~unif(14%, 18%)*pp, ~unif(25%, 35%)*pp, and ~unif(45%, 55%)*pp, respectively. Inspection cost of Tier k returns at the hybrid facility and the recovery center are modeled as a function average refurbishing cost of that tier, ~unif (8%, 12%) * average refurbishing cost of Tier k returns at the manufacturer. The demands at the retailers are uniformly distributed between 500 and 700 units.

Product return percentage of 30% is considered in this example. Out of the product returns, 50% are Tier 1 returns, 30% are Tier 2 returns, 10% are Tier 3 returns, and 10% are Tier 4 returns. Tier 1 and Tier 2 returns account for more than 90% of the total returns based on a case study published in Ovchinnikov (2011) for a major wireless carrier in North America. We assume 80% of the returns to be Tier 1 and Tier 2 returns. Tier 3 and Tier 4 returns are assumed to be 10% each in order to discuss their effect on the supply chain network design.

We assume that the refurbished product quality in the market is uncertain and falls between $[q - i = 0.30, q + i = 0.7]$. The corresponding perceived value falls between $[v(q - i) = 0.55, v(q + i) = 0.84]$ (see data discussed in Chapter 3, Section 3.4). The acceptance rate of low-end customers is calculated using Equation (3.1) and that of high-end customers is modeled using an inverted U-shaped function, given in the Equation (3.2). In this example, the discount offered for refurbished products is initially set at 25%. Customer acceptance rate for low-end customers it is calculated using Equation (3.1) as 0.3103 (31.03%) and for high-end customers is calculated using Equation (3.2) as 0.1364 (13.64%). The discount rate will be varied as part of the sensitivity analysis in Section 4.4.

The data for this example is coded and generated using Microsoft Visual C++ 6.0. The mathematical model is coded and solved using a commercial optimization software package. The model for this example has 11,723 variables (11,643 continuous variables and 80 binary variables) and 808 constraints. The model took approximately 12 seconds to solve to optimality. The optimal profit achieved by the model for this example is $12,295,957.

The total demand across all the retailers is 60,028 units. Retailer returns in the model are calculated using the left-hand side of the Equation (4.7), that is, demand of the retailer times the return percentage and the percentage of Tier k returns, summed over all retailers, and return tiers. For this illustrative example, retailer returns are 17,837, out of which 8,958 (50%) are Tier 1, 5363 (30%) are Tier 2, 1758 (10%) are Tier 3, and 1758 (10%) are Tier 4 returns.

Total acceptance for refurbished items is calculated using the left-hand side of the Equation (4.6), that is, demand of the retailer times the return percentage of high/low-end customers and their acceptance rates for the discount offered, summed over all retailers. The total acceptance for refurbished items is 15,449 units, based on the acceptance rates of low-end and high-end customers in this illustrative example. Since the acceptance rate is less than the returns, some products are disposed-off at the retailers as these refurbished products do not have demand. In the optimal solution, all the Tier 1 returns, and Tier 2 returns are refurbished; 630 units of Tier 3, and 1,758 units of Tier

4 returns are disposed-off at the retailers. We can observe that higher-tier returns are disposed-off because their refurbishing costs are higher. Thus, it is important to consider different categories of return and customer buying behavior in our model to avoid sub-optimal solutions.

The optimal network design is as follows: Suppliers 1, 3, 4, 5, 6, 7, 8, 9, and 10 are selected to supply raw materials to the manufacturing plants. All the five manufacturing plants are used. Plants 1, 2, 3, and 5 are used in new product production and Plants 1, 4, and 5 are used for refurbishing the return products. Warehouses and recovery centers are not used for distribution in the optimal solution. Hybrid facilities 2, 6, and 8, each of size 3, are opened and utilized to 100% of their capacity. The retailers' demands are satisfied from Hybrid facilities 2, 6, and 8 in the forward channel and the returns at the retailers are shipped to Hybrid facilities 2, 6, and 8 in the return channel. Interested readers can refer to Pazhani (2014) for a detailed discussion on illustrative example and comparison between Base Model and Extended Base Model solutions.

4.3.1 Comparison between Base Model and Extended Base Model Solutions

In this section, we compare the optimal solutions obtained from the Base Model and the Extended Base Model. The Extended Base Model considers different tiers of returns and customer categories as additions to the Base Model. In the realistic case (Chapter 2), customer acceptance rate was considered as 30%. The illustrative example in Chapter 4 considers a customer acceptance rate of 0.25813 (0.3103 × 0.7 + 0.1364 × 0.3). In order to have a fair comparison between the two solutions, we set the customer acceptance rate as 0.25813 in the Base Model. Now, the total customer acceptance rate is 0.25813 (25.813%) in the Base Model also.

The total profit in the Base Model is $11,769,597 and $12,295,957 in the Extended Base Model. The Extended Base Model yields 4.47% higher profit than the Base Model solution. We will now discuss the differences in the optimal solutions from the Base Model and Extended Base Model: Suppliers 1, 3, 4, 5, 6, 7, 8, 9, and 10 are selected to supply raw materials in both the Base Model and the Extended Base Model. However, the quantities of raw materials purchased from the selected set of suppliers differ. Suppliers 3, 4, 10 were allocated to supply 2,975 units, 1,982 units, 1,708 units, respectively, in the Base Model, as opposed to 5,078 units, 437 units, 1,150 units, respectively, in the Extended Base Model. Manufacturing and refurbishing plans at the plants are different between the models. In the Base Model, 9,357 units in Plant 1, 1,707 units in Plant 2, 16,381 units in Plant 3, and 17,134 units in Plant 4 of new products are produced. In the Extended Base Model, 11,460 units in Plant 1, 1,150 units in Plant 2, 16,381 units in Plant 3, and 15,589 units in Plant 4 of new products are produced. 6,073 units in Plant 1, 2,927 units in Plant 4, and 6,449 units in Plant 5 of the products are refurbished in the Base Model solution. The refurbishing plan in the Extended Base Model is more detailed. Plant 1 refurbishes 3,970 units, out of which 3,096 are Tier 1 returns and 874 are Tier 2 returns; Plant 4 refurbishes 3,485 units, out of which all are Tier 1 returns; and Plant 5 refurbishes 7,995 units, out of which 5,474 are Tier 1 returns, 2,267 are Tier 2 returns, and 254 are Tier 3 returns. Hybrid facilities 2, 6, and 8, each of size 3 are opened in the optimal solutions of both the models. Warehouses and recovery centers are not opened in either model. We observe that the network designs (supplier selection, selection of hybrid facilities, warehouses, and recovery centers) in both models are the same. However, the quantity allocated to suppliers, manufacturing, and refurbishing plans at the plants, and distribution plan at the hybrid facilities differ between the models, a situation that impacts the total profit of the supply chain.

Alternatively, the solution from the Extended Base Model is more detailed. It might help supply chain managers plan their refurbishing activities better. For example, Plant 4 can only hold inventories of items for refurbishing common defects in Tier 1 returns. Due to the lack of acceptance for refurbished products, the optimal solution from both models recommend to dispose-off the same amount of returned products (2,388 units). Base Model shows that 2,388 units need to be disposed-off at the retailers. The Extended Base Model shows that 630 units of Tier 3 and 1,758 units of Tier 4 returns need to be disposed-off. Supply chain managers can use the detailed solutions to efficiently plan disposal strategies for returns at the retailers. The solution from the Extended Base Model might also assist decision-makers in exploring other recycling activities for the disposed products, such as selling them in a secondary market. The distribution of new products, refurbished products, and returned products in the hybrid facilities also differ between the two models. Quantity allocated to the suppliers is also an important decision for price negotiation and contracts. This comparison shows that it is important to explicitly consider the different tiers of returns and their refurbishing costs in the model. In the sensitivity analysis section, we will show that the customer categories have to be incorporated in the model in order to determine the optimal discount levels for the refurbished products.

4.4 SENSITIVITY ANALYSIS – EXTENDED BASE MODEL

In this section, we perform a sensitivity analysis with the following purposes: (1) to study the effect of varying customer return rate, percentages of low-end and high-end customers, and discounts offered for refurbishing products on the total profit of the supply chain. In this analysis, we fix the refurbished product quality in the market. A total of 2,184 cases are analyzed; (2) to study the effect of varying product quality in the market, and discounts offered for refurbishing products on the total profit and design of the supply chain network. In this analysis, we fix the customer return rate and the mix of low-end and high-end customers. A total of 52 cases are analyzed.

4.4.1 Varying Customer Return Rate, Percentage of Low-End and High-End Customers, and Discounts Offered for Refurbishing Products

In this analysis, we fix product quality in the market to be [$q - i = 0.30$, $q + i = 0.7$] and hence, the corresponding perceived value falls between [$v(q-i) = 0.55, v(q+i) = 0.84$]. Thus, the customer acceptance rate for low-end customers is calculated, using Equation (3.1), as [$0.84 - (rp/np)]/(0.84 - 0.55)$. The acceptance rate of high-end customers, is calculated using Equation (3.2): $\alpha_2 = 0.008 - (1.227 \times \theta) + (13.45 \times \theta^2) - (31.12 \times \theta^3) + (20.68 \times \theta^4)$, will be fixed for a given discount rate θ, which is equal to $(1 - (rp/np))$. This function is generated based on the hypothetical data provided in Chapter 3, Section 3.5.

We consider 4 levels for the split between low-end and high-end customers: 50%-50%, 60%-40%, 70%-30%, 80%-20%. Twenty-one levels are considered for the customer return rate parameter, which is varied between 0.1 and 0.50 in steps of 0.02. Discounts offered for refurbished products are varied between 0.04 and 0.54 in increments of 0.02 (26 levels). A total of $4 \times 21 \times 26$, equal to 2,184, cases are modeled and optimal solutions are obtained.

The product quality range is fixed between 0.3 and 0.7. So, the acceptance rate of low-end customers depends only on the level of discount offered. In practice, acceptance rates of high-end customers are usually less than those of the low-end customers. Since the low-end customer acceptance rate is 0% for discounts up to 16%, we set the customer acceptance rate for high-end customers to be also at 0% up to a 16% discount level. For discount levels higher than 16%, the customer acceptance rate curve for high-end customers is calculated using Equation (3.2). Note that the customer acceptance rate for high-end customers declines after the 40% discount level.

As low-end customer acceptance rate linearly increases with increase in discount rate and as the high-end customer acceptance rate is an inverted U-shaped function with respect to the discount rate, the spread between the acceptance rates of low-end and high-end customers diverge as discount rate increases. More empirical research is needed to validate the linear relation between the customer acceptance rates and discount levels for low-end customers. However, literature shows that a linear function is fairly accurate.

Table 4.1 shows the average profit of the supply chain under different discount levels and mix of low-end/high-end customers.

First, from the analysis, we observe that the profit also follows an inverted U-shaped function for a given customer mix (Table 4.1). At low discount levels, the customer acceptance rates for both low-end and high-end customers

TABLE 4.1 Average profit of the supply chain under different discount levels and customer mix

DISCOUNT OFFERED	% OF LOW-END/HIGH-END CUSTOMERS			
	50/50	60/40	70/30	80/20
4%	9,812,067	9,812,067	9,812,067	9,812,067
8%	9,812,067	9,812,067	9,812,067	9,812,067
12%	9,812,067	9,812,067	9,812,067	9,812,067
16%	9,812,067	9,812,067	9,812,067	9,812,067
20%	11,425,508	11,500,510	11,577,129	11,646,229
24%	12,104,365	12,242,884	12,352,234	12,465,240
28%	12,243,640	12,356,623	12,466,266	12,537,041
32%	11,964,868	12,035,739	12,068,242	12,088,413
36%	11,441,967	11,454,619	11,455,561	11,455,780
40%	10,819,979	10,820,557	10,820,683	10,820,685
44%	10,208,093	1,020,8239	10,207,979	10,208,239
48%	9,820,396	9,820,396	9,820,396	9,820,396
54%	9,812,067	9,812,067	9,812,067	9,812,067

are very low. Hence, more of the returned products will be disposed-off at the retailers (because of lack of demand for refurbished products). The profit then increases as the discount rate increases, and more customers (low- and high-end) are willing to buy refurbished products. After a certain level of discount (at higher discount levels), the acceptance rate decreases for high-end customers and the profit of the supply chain decreases because it is not profitable to offer more than that certain level of discount on the refurbished products.

The analysis also shows that the customer mix has a significant impact on the profit of the supply chain. At a given discount level, we observe that the total profit increases with increases in the percentage of the low-end customers (Table 4.1). This is due to the fact that the acceptance rate of high-end customers is less than that of the low-end customers. This can also be due to the inverted U-shaped acceptance behavior of high-end customers. For example, assume that the discount offered is 20% and out of 100 customers, 50 are high-end and 50 are low-end customers. If the customer acceptance rate of low-end customers is 20%, 10 low-end customers are willing to accept refurbished products. With 20% discount, say, 10% of high-end customers are willing to accept refurbished products, that is, 5 customers. Assume that the total returns are 20 units, then only 15 (10 + 5) products can be refurbished and sold (if it is profitable to the supply chain). When there are 20 high-end and 80 low-end customers, 18 products (16 + 2) can be refurbished.

4.4.2 Varying Refurbished Product Quality in the Market and Discounts Offered for Refurbishing Products

In this analysis, we fix the customer return rate to 0.32, and the percentage of low-end and high-end customers to be 70% and 30%, respectively. Product quality in the supply chain affects the customer acceptance rate of low-end customers. However, the customer acceptance rate of high-end customers is unaffected by product quality as high-end customers judge the quality of the refurbished product based on product price, namely, the discount rate. Thus, in this analysis, we also fix the customer acceptance rate of high-end customers.

Let us analyze the effect of varying quality perception of refurbished products in the market and discount levels on the total profit and network design of the supply chain. We will consider two quality scenarios or two levels: (1) the product quality in the market falls between $[q - i = 0.30, q + i = 0.70]$

and (2) the product quality falls between $[q - i = 0.40, q + i = 0.80]$. Discounts offered for refurbished products are varied between 0.04 and 0.54 in increments of 0.02 (26 levels). A total of $2 \times 26 = 52$ cases have been modeled and optimal solutions are obtained.

The scenario with $[q - i = 0.30, q + i = 0.70]$ will be called Scenario 1 (lower quality) and the scenario with $[q - i = 0.40, q + i = 0.80]$ will be called Scenario 2 (higher quality). Using the hypothetical data discussed in Chapter 3 (refer to Section 3.4), the corresponding perceived product values for Scenarios 1 and 2 fall between $[v(q - i) = 0.55, v(q + i) = 0.84]$ and $[v(q - i) = 0.63, v(q + i) = 0.89]$, respectively. The low-end customer acceptance rate for Scenario 1 is $[0.84 - (rp/np)]/(0.84 - 0.55)$ and that for Scenario 2 is $[0.89 - (rp/np)]/(0.89 - 0.63)$, where $(rp/np) = 1 - \theta$ and θ is the discount rate offered.

Under Scenario 2, the low-end customer acceptance rate is 0% for discounts up to 11%. Hence, we set the customer acceptance rate for high-end customers also at 0% up to an 11% discount level. For discount levels higher than 11%, the customer acceptance rate for high-end customers is calculated using Equation (3.2). The acceptance rate for high-end customers for Scenario 1 is discussed in Section 4.4.1. Note that in both scenarios, the acceptance rates of low-end customers increase with increases in discount rates, while for high-end customers, they follow the inverted U-shaped curve. For low-end customers, the acceptance rate is higher when the quality is higher. High-end customers are generally not affected by change in quality. Table 4.2 shows the average profit of the supply chain under different discount levels in Scenarios 1 and 2.

4.4.2.1 Changes in optimal discount levels due to changes in return product quality in the market

As observed from Table 4.2, the maximum profit in Scenario 1 occurs at a discount rate of 24% and in Scenario 2 at a discount rate of 20%. Thus, the changes in the quality perception of the returned products in the market, which affect the acceptance rate of low-end customers, lead to changes in the profit of the supply chain. Customer behavior of low- and high-end customers is an important factor in setting the discount rates for the refurbished products in the supply chain. Given the empirical data for the acceptance rate for high-end customers, the optimal level of discount in the supply chain has to be decided based on the quality levels (which affects the low-end customer acceptance rate). Setting a discount level without considering it will lead to suboptimal solutions.

TABLE 4.2 Profit of the supply chain under different discount levels in Scenario 1 and Scenario 2

DISCOUNT OFFERED	SCENARIO 1 (LOWER QUALITY)	SCENARIO 2 (HIGHER QUALITY)
4%	9,812,067	9,812,067
6%	9,812,067	9,812,067
8%	9,812,067	9,812,067
10%	9,812,067	9,812,067
12%	9,812,067	10,346,829
14%	9,812,067	11,376,321
16%	9,812,067	12,284,068
18%	10,875,562	12,865,602
20%	11,606,260	13,183,122
22%	12,135,492	12,946,415
24%	12,379,815	12,641,228
26%	12,324,965	12,334,039
28%	12,031,577	12,031,577
30%	11,750,496	11,750,496
32%	11,475,617	11,475,617
34%	11,203,075	11,203,075
36%	10,925,861	10,931,185
38%	10,659,535	10,659,535
40%	10,443,586	10,443,586
42%	10,267,497	10,267,497
44%	10,097,757	10,097,757
46%	9,928,017	9,928,017
48%	9,820,446	9,820,446
50%	9,812,067	9,812,067
52%	9,812,067	9,812,067
54%	9,812,067	9,812,067

4.4.2.2   Changes in optimal network design due to changes in return product quality in the market

We will now discuss the differences in network design corresponding to the maximum profits of Scenarios 1 and 2. The maximum profit in Scenario 1 is $12,379,815 (corresponds to 24% discount rate) and Scenario 2 is $13,183,122 (corresponds to 20% discount rate – highlighted in Table 4.2). In

Scenario 1, Suppliers 1, 3, 5, 6, 7, 8, 9, and 17 are selected to supply raw materials. Whereas in Scenario 2, Suppliers 1, 3, 5, 6, 7, 8, 9, and 10 are selected to supply raw materials. We also see that the manufacturing and refurbishing plans are different between the two scenarios. In Scenario 1, Plants 1, 3, and 5 are used to produce new products and Plants 1, 4, 5 are used to refurbish products. In Scenario 2, Plants 1, 2, 3, 5 are used to produce new products and Plants 1, 4, 5 are used to refurbish products. In Scenario 1, 46,206 new products are produced, and 13,822 units are refurbished. In Scenario 2, 44,015 new products are produced, and 16,013 units are refurbished. More refurbished products are made when the quality is higher (Scenario 2). Hybrid facilities 6 and 7, each of size 3, are opened in Scenario 1. Hybrid facilities 2, 6, and 8, each of size 3, are opened in Scenario 2. Warehouse 14 of size 3 is opened in Scenario 1. A total of 22,118 units of products (15,398 of new products and 6,720 units of refurbished products) are distributed to retailers. No warehouses are opened in Scenario 2.

Between the optimal solutions of Scenarios 1 and 2, we see that there is a difference in location selection for hybrid facilities and warehouses, supplier selection decisions, manufacturing and refurbishing plans at the plants, and distribution of products in the supply chain. Location and supplier selection in a supply chain are strategic-level decisions and directly affect the profitability of the supply chain.

In summary, the following are the inferences from the sensitivity analysis: (1) for a given customer mix, the supply chain profit follows an inverted U-shaped function with respect to the discount levels. This is because the customer acceptance rate for refurbished products is low at lower discount levels. For higher discount levels, the customer acceptance rate for refurbished products is low for high-end customers but high for low-end customers; (2) discount levels have a significant impact on the profit of the supply chain; (3) given a discount level, the total profit increases with increases in the percentage of the low-end customers. This is because of two reasons: inverted U-shaped behavior of high-end customers, and the fact that the acceptance rate of low-end customers increases with higher discount rates; (4) optimal discount level for the refurbished products has to be determined based on the customer acceptance rate. Particularly, our analysis shows that the quality of the returned products is an important factor in determining the optimal discount rate for the refurbished products in the supply chain; and (5) network design varies with varying customer acceptance behavior for refurbished products. We show that the change in return product quality in the market (which affects the low-end customer acceptance rate) has an impact on the supply chain network design. In the illustrative example, the network designs between Scenario 1 and 2 vary in terms of the selection of the hybrid facilities and the warehouses, supplier selection decisions, manufacturing and refurbishing plans at the plants, and distribution of products in the supply chain.

4.5 MANAGERIAL IMPLICATIONS

In this section, we will discuss the managerial implications of the model and the analysis presented in this chapter. This framework proposed in this chapter allows managers to design more realistic CLSC networks by considering the different tiers of returns and customer behavior. In particular, the model determines whether to refurbish or dispose-off returns with major defects based on the total supply chain profitability. Compared to the Base Model developed in Chapter 2, the Extended Base Model gives a design solution that is more comprehensive. The model provides a detailed refurbishing plan at the manufacturing plants so that the managers can plan their refurbishing activities, that is, manpower, equipment planning, parts inventories specific to the type of returns flowing into that plants, etc. The company can devise a disposal strategy based on the optimal solution from the model, that is, the managers can decide on the percentage of returns (in each tier) to be scrapped or sold to a secondary market at the retailer, so that it is profitable.

The analysis in this chapter helps us understand the sensitivity of varying customer mix and how their buying behavior affects the supply chain profit and the network design. Using the framework, we analyze different discount levels for refurbished products to design their supply chain optimally. Our sensitivity analysis shows that the optimal discount level has to be set based on the customer behavior.

In summary, the Extended Base Model will help managers make better decisions by considering more realistic factors, such as different tiers of returns and customer behavior in buying refurbished products in their supply chain network design.

The following are the challenges in building this model: (1) Categories of the returns in the model and their corresponding refurbishing cost: data related to returned products at the retailers along with its associated defect is needed. A detailed statistical analysis on the historical returns can be performed to create significantly different groups of return categories. Refurbishing cost can be estimated based on the required repairs (based on the defect data); (2) Percentage of customer categories and their buying behavior. These parameters often require detailed sales data analysis. In practice, customer categories can be more than low-end and high-end. It depends on how different their buying behaviors are. If the company is starting to explore refurbishing, it can also analyze similar products in the market. However, ignoring these parameters can lead to sub-optimal network design.

Companies can use this modeling framework to optimally design or redesign their CLSC network, given their estimation on return categories parameters, their refurbishing cost, and customer categories and their buying behavior is fairly accurate.

4.6 SUMMARY

In the Base Model in Chapter 2, we showed the gain in adopting recycling and refurbishing practices in the supply chain. We also observed that the return parameters, customer acceptance behavior, and refurbishing cost have a significant impact on the supply chain profitability and the network design. In this chapter, we addressed these questions: (a) is it profitable to design your CLSC incorporating these parameters? and (b) what is your refurbishing strategy in your supply chain depending on the return parameters? The condition of the returned products and the cost of refurbishing them vary significantly. In this chapter, we extended the Base Model of Chapter 2 by considering different tiers of returns and customer buying behavior in purchasing refurbishing products. We extended the modeling framework and developed a MILP model for the problem. In the MILP model, we considered two customer categories (low-end and high-end customers), and four tiers of returns based on the quality and condition of the products. We considered the refurbishing cost based on the tier of return and modeled the acceptance rates of low-end and high-end customers as a function of the discount rates offered to the refurbished products. The model determined the optimal location of the facilities and the distribution of new, refurbished, and returned products in the CLSC network.

We analyzed the Extended Base Model with an illustrative example. We compared the optimal solutions obtained from the Extended Base Model with the results of an equivalent Base Model. The comparison showed that the Extended Base Model would help the supply chain managers to make more informed decisions. It also showed that incorporating different tiers of returns and their refurbishing costs had an impact on the total profit of the supply chain. In the Extended Base Model, the profit was 4.5% higher than that of the equivalent Base Model.

The discount level for the refurbished products is a critical decision that significantly affects the profit of the supply chain. The Extended Base Model analyzed the impact of the low-end and high-end customers and the discount levels for refurbished products on the total profit of the supply chain. Sensitivity analysis was performed by varying customer return rate, proportion of low-end and high-end customers, discounts offered for the refurbished products, and the customer acceptance rate for refurbished products. The results showed that the proportion of low- and high-end customers had a significant impact on the profit. Also, we showed that the discount levels should be set based on the customer acceptance rates for refurbished products. We also observed that the network design changed with varying customer acceptance behavior. Therefore, these aspects should be modeled to avoid loss of potential profits in the CLSC network.

It can be very challenging to accurately model customer behavior and return rate, even with huge volumes of good data and modeling techniques. However, incorporating these parameters is critical for the profitability of the supply chain. Chapter 5 proposes a robust optimization modeling framework by considering customer behavior and return rate aspects as uncertain parameters.

4.7 MULTI-OBJECTIVE EXTENSIONS

As discussed in Chapter 2, the model framework in this chapter can also be extended using multiple objectives. A multi-objective extension of the Extended Base Model can be found in Pazhani (2016). The bi-criteria model in the published chapter focuses on designing an optimal CLSC network with the objectives of maximizing the supply chain profit and minimizing the energy usage at the warehousing facilities and energy consumed during transportation.

REFERENCES

Ovchinnikov, A. (2011), 'Revenue and cost management for remanufactured products', *Production and Operations Management* **20**(6), 824–840.

Pazhani (2014). Design of Closed-Loop Supply Chain Networks, The Pennsylvania State University, https://etda.libraries.psu.edu/files/final_submissions/9801

Pazhani, S. (2016). 'A Bi-criteria model for closed-loop supply chain network design incorporating customer behavior', In *Multiple Criteria Decision Making in Supply Chain Management*, pp. 243–284. CRC Press, Boca Raton, FL, USA.

Robust Network Design Incorporating Uncertainties in Demand and Return Parameters

5

5.1 UNCERTAINTY IN CLOSED-LOOP SUPPLY CHAIN

In Chapter 4, we analyzed the CLSC network design problem by incorporating customer buying behavior toward refurbishing products and refurbishing cost based on the return product quality. We observed that these parameters have a significant impact on the profit, discount levels for refurbished products, and on the network design of the supply chain. Thus, incorporating these parameters are critical for the profitability of the supply chain. The following questions arose from the analysis: (a) is quality data available to ascertain these parameters?; (b) in particular, can we accurately model customer categories and behavior? Therefore, in this chapter, we focus on designing the supply chain by incorporating the fuzziness/uncertainty in these modeling parameters. We develop a *robust optimization* modeling framework to incorporate uncertainties and to design a robust CLSC design. Section 5.1.1 gives a quick overview of robust optimization models and how they can be used to incorporate uncertain parameters.

5.1.1 Introduction to Robust Optimization Models

Classical models that consider uncertainty in the input parameters include *mean-risk model* (Markowitz, 1952), *recourse model* (Dantzig, 1955), and *chance-constrained model* (Charnes *et al.*, 1958).

The mean-risk model was first proposed for the portfolio selection problem. Markowitz suggested that investors consider portfolios that maximize the expected returns and minimize the variance (risk) of the return. Chance-constrained model, introduced by Charnes *et al.* (1958), replaces the deterministic constraints in the traditional mathematical programming with probabilistic constraints, where some or all input parameters are random and the constraints are required to hold with at least some level of reliability < 1. In order to solve the problem using the chance-constrained model, the probability distribution of the uncertain parameters have to be known. Recourse models are two-stage (or multi-stage) mathematical models where decisions made in the first stage are corrected in the second stage (or later stages) based on the realizations of the problem's random components. The first-stage decision variables can be regarded as proactive and are often associated with planning issues, such as capacity expansion or aggregate production planning. Second-stage decision variables can be regarded as reactive and are often associated with operating decisions. These second-stage decisions allow to model a response to the observed outcome, which constitutes the recourse (Sen and Higle, 1999). Recourse in these models is done at the expense of some estimated recourse cost. For example, in production and inventory systems, the first-stage decision might correspond to production quantities, and demand might be modeled using random variables. When demand exceeds the amount produced, policy may dictate that customer demand be backlogged at some cost. In this case, the precise estimation of the backlog cost plays an important role in the model.

The concept of 'Robust optimization' emerged from the dissatisfaction with the limitations of the classical models (Greenberg and Morrison, 2008). Firstly, Gupta and Rosenhead (1968) introduced the notion of *flexibility*, which refers to how many and what recourses are available in each feasible alternative. Their definition of a robust solution is based on the flexibility it allows after the uncertain values become known. Mulvey *et al.* (1995) introduced the concept of model and solution robustness using a penalty function like the recourse model. Their robust model penalizes the risk of the uncertain parameters as in the Markowitz model. In this chapter, we use the robust optimization model proposed by Mulvey *et al.* (1995) with the objective of maximizing the supply chain profit and penalizing the variation of the profit across the scenarios, that is, minimizing the variance across all the scenarios.

Other robust optimization models include worst-case hedge model, simple case of interval uncertainty model, minimax regret model, and uncertainty sets model. A detailed review of these models along with the applications is given in Chapter 14 of *Operations Research and Management Science Handbook* edited by Ravindran (2008).

One of the methodologies to represent uncertainty of input parameters in stochastic programming is the scenario-based approach (Gupta and Maranas, 2003). In the scenario-based approach, a set of discrete future scenarios is generated, where each scenario describes discrete values for some uncertain parameter(s) and is associated with a probability of occurrence determined by a decision maker (Solo, 2009). Mulvey *et al.* (1995), Alonso-Ayuso *et al.* (2003), Guillén *et al.* (2005), Santoso *et al.* (2005), and Leung *et al.* (2006) present some scenario-based approaches to stochastic supply chain problems.

A scenario-based robust model has two types of variables: (1) structural or design variables, whose optimal values are not dependent upon the realization of uncertain input parameters; (2) control variables, whose optimal values depend upon the realization of uncertain parameters, as well as the optimal values of the design variables (Mulvey *et al.*, 1995). It should be noted that the values of the design variables (typically the location selection variables) cannot be adjusted based on the realization of the uncertain data. A set of scenarios is used to reflect the different values for the uncertain input data in the model.

Let a and b be the set of design and control decision variables, respectively. Let c be the cost matrix associated with the design variables and d be the cost matrix associated with the control variables. The objective function for the deterministic problem is in the form $(c^T a + d^T b)$. In the presence of uncertainty, the objective function is defined over each of the scenarios and is in the form ($c^T a + d^T_x b$) for Scenario x which belongs to the set X. Let pr_x be the probability of occurrence of Scenario x and *profit$_x$* be the profit of the supply chain in Scenario x. In a general stochastic linear programming model, the objective is to maximize the expected value of *profit$_x$* over all Scenarios x in $X = \{$sum x in X ($pr_x \cdot$ *profit$_x$*)$\}$. This model only maximizes the expected value across all possible scenarios leaving out the potential variability in value (across scenarios) associated with the uncertain parameter(s). Mulvey *et al.* (1995) proposed a mean/variance approach as a technique for mitigating the risk associated with one or more uncertain input parameters. The authors proposed a revised function considering both the expected profit and weighted variance, with λ as a positive weight for the variance term. As the value of the positive weight term increases, the solution becomes less sensitive to changes in the data defined by the scenarios (Solo, 2009). It enables the robust optimization model to account for the decision maker's preferences toward risk (Mulvey *et al.*, 1995). With variability under control, minimal adjustment to the control variables will be required when the weighted variance term is used in the objective function (Solo, 2009). However, the variance term introduces a quadratic term, which makes the model non-linear. To remove this quadratic term, Yu and Li (2000) proposed an alternative formulation, by considering the absolute value of the deviation of the profit of Scenario x from the expected profit across all the scenarios. Interested authors are referred to Yu and Li (2000) for detailed mathematical discussion.

Most of the scenario-based optimization models in CLSC design consider the objective of minimizing the expected cost or maximizing the expected profit. With such objectives, the optimal design will be for the worst possible scenario (i.e., to accommodate highest demand and returns). However, the variability of the uncertainty parameters (across scenarios) has not been considered in any of the studies. In this chapter, the uncertain parameters are modeled using discrete scenarios with associated probabilities. A scenario-based robust optimization model is developed to maximize the expected profit of the supply chain along with minimizing the variability of the profit across the scenarios. Minimizing the variability is necessary to mitigate the risk associated with the uncertain input parameters in the supply chain.

The remainder of the chapter is organized as follows: Section 5.2 presents the robust optimization framework for the Extended Base Model. In Section 5.3, we present an illustrative example to show the application of the Extended Base Model. In Section 5.4, we present an analysis of the robust model solution along with comparison of deterministic and robust designs. Section 5.5 discusses the advantages of the robust optimization. In Section 5.6, we will study the effect of varying discounts offered for the refurbished products on the total profit and design of the supply chain network. Under optimal discount setting, a comparison of deterministic and robust designs is also presented. Section 5.7 presents the conclusions of the chapter.

5.2 PROBLEM DESCRIPTION AND THE MODEL FRAMEWORK

This section presents the robust optimization framework for the Extended Base Model. The problem description is the same as discussed in Chapter 2 and extended in Chapter 4. In this chapter, we incorporate uncertainty in three key input parameters: demand at the retailers, return rate at the retailers, percent of low-end and high-end customers willing to buy refurbished items (i.e., customer buying behavior). The framework is generic and flexible to incorporate other uncertain parameters as well. The model considers X discrete scenarios and one or more uncertain parameters are varied for each scenario. Let pr_x be the probability of occurrence of Scenario x in X, where X is the set of scenarios, and the sum of pr_x across all Scenarios is equal to 1. Let d_{cx} be the demand for products at retailer c in Scenario x, γ_x be the percent of demand returned at retailer in Scenario x, α_{1x} and α_{2x} be the percent of low-end and high-end customers willing to buy refurbished items in Scenario x, respectively. Let λ be the constant for the weighted variance term.

All the cost components (purchasing, manufacturing, refurbishing, transportation, and fixed costs) are the same across all scenarios. Note that this model considers four tiers of returns (Tiers 1, 2, 3, 4), like Chapter 4. However, the model can be easily modified to incorporate uncertainty in refurbishing costs or any other cost components. The capacities of the facilities do not change across scenarios. All the decision variables, except the facility location selection binary variables, are extended and defined for each scenario: QSM_{smx} is the quantity of raw materials purchased from Supplier s by Plant m in Scenario x, QMW_{mwx} is the quantity of new products transported from Plant m to Warehouse w in Scenario x, $RQMW_{mwx}$ is the quantity of refurbished products transported from Plant m to Warehouse w in Scenario x, QMH_{mhx} is the quantity of new products transported from Plant m to Hybrid facility h in Scenario x, $RQMH_{mhx}$ is the quantity of refurbished products transported from Plant m to Hybrid facility h in Scenario x, QWC_{wcx} and $RQWC_{wcx}$ is quantity of new products and refurbished products transported from Warehouse w to retailer c in Scenario x, QHC_{hcx} is the quantity of new products transported from Hybrid facility h to retailer c in Scenario x, $RQHC_{hcx}$ is the quantity of refurbished products transported from Hybrid facility h to retailer c in Scenario x, $RQCH_{kchx}$ is the quantity of returned products in Tier k transported from retailer c to Hybrid facility h in Scenario x, $RQCR_{kcrx}$ is the quantity of returned products in Tier k transported from retailer c to recovery center r in Scenario x, $RQHM_{khmx}$ is the quantity of returned products in Tier k transported from Hybrid facility h to Plant m in Scenario x, $RQRM_{krmx}$ is the quantity of returned products in Tier k transported from recovery center r to Plant m in Scenario x, $DISP_{kcx}$ is the quantity of return products in Tier k disposed at retailer c in Scenario x, δ_w^l is the location selection binary variable for Warehouse w with capacity l, η_h^l is the location selection binary variable for Hybrid facility h with capacity l, and ξ_r is the location selection for recovery center r. We introduce two new variables in this model: let UN_{cx} be the unsatisfied demand at retailer c in Scenario x, and Θ_x be the slack variable for the weighted variance term for Scenario x.

5.2.1 The Model

The framework of the optimization model is as follows: the objective of the model is to maximize the expected total profit of the CLSC across all scenarios and minimize mean absolute deviation (variance) of profit across all scenarios. Interested readers can refer to Pazhani (2014) for a detailed mathematical model.

The proposed Robust Model can be accessed from https://www. routledge.com/Design-and-Analysis-of-Closed-Loop-Supply-Chain-Networks/Pazhani/p/book/9780367537494 (see Robust Model section). All the equation numbers are references from the web link.

Constraint (5.2) calculates the slack variable used in the variance term. If the left-hand side of the constraint is greater than 0, that is, if the profit of Scenario x is greater than or equal to the expected profit of the supply chain across all scenarios, then the slack variable is 0 in the optimal solution. If the left-hand side of the constraint is less than 0, that is, if the profit of Scenario x is less than the expected profit of the supply chain across all scenarios, then the slack variable for Scenario x is the deviation from the expected profit across all scenarios and is positive in the optimal solution. Constraint set (5.3) represents the demand constraints. The total quantity of products (new and refurbished) flowing into retailer c plus the unsatisfied demand in retailer c should be equal to the demand at that retailer in Scenario x.

5.3 ANALYSIS OF THE ILLUSTRATIVE EXAMPLE

We consider the same four-stage supply chain network as discussed in Chapter 2, Section 2.2. The cost parameters are modeled as a function of the new product price similar to the realistic case discussed in Chapter 2 (see Section 2.6). The capacities of the suppliers and plants, capacities and fixed cost of the warehouses, hybrid facilities, and the recovery centers are also the same as given in Chapter 2 (see Section 2.4 and Tables 2.1, 2.2, 2.3, and 2.4).

As discussed in Chapters 2 and 4, the cost parameters are modeled as a function of the new product price. Refurbishing cost and inspection cost of returned products are set based on their tier of return, that is, higher the tier of return, higher is the cost of refurbishing and inspection (See Section 4.3). Initially, we fix the discount rate in the scenarios. We will then vary the discount rate and determine the best discount rate for the illustrative example. We also fix the customer mix to 70% low-end and 30% high-end customers. We consider uncertainty in demand, quantity of returns, perceived quality of refurbished products in the markets and the corresponding customer acceptance rate for low-end customers.

5.3.1 Design of the Scenarios for the Illustrative Example

In this section, we will discuss how the scenarios are created for the illustrative example.

Demand: We consider four possible economic scenarios in the market: *Strong*, *Good*, *Fair*, and *Weak*. Forecasts of the demand at the retailers in each of the economic scenarios are assumed to follow uniform distributions as follows: Strong demand scenario with demand ranging ~unif (700, 900), Good demand scenario with demand ranging ~unif(500, 700), Fair demand scenario with demand ranging ~unif(400, 500), Weak demand scenario with demand ranging ~unif(300, 400).

Return rate and quality of the new product: 95% of commercial returns happen within 90 day of sales if they do not meet customers' expectations for some reason, customer remorse, or if the customer finds a better price. Shipping damages and product defects account for the remaining 5%. Thus, the perception of the new product quality in the market is one of the key variables in determining the product return rate. We assume a linear inverse relationship between the product quality and the return rate, that is, as product quality increases, return rate decreases. Assuming 100% represents perfect quality, we will consider three different quality levels for the new product in the market: Low (65%), Medium (80%), and High (95%). We model the return-rate scenarios based on three different product quality levels and refer to them as low quality, medium quality, and high quality. In this example, the corresponding return rates for Low, Medium, and High quality levels are hypothetically assumed to be 0.5, 0.3, and 0.2 (higher the product quality, lower the return rate), respectively. Note that these values are hypothetical, and an empirical study must be performed to determine the return rates as a function of quality for the system of interest.

Quality perception of the refurbished products and acceptance rates of customers: Uncertainty about the condition of the returned products (why and when it was returned, what went wrong), and details of refurbishing carried out on the returned products are some of the factors that influence the customer's quality perception of the refurbished products. Quality of the new product in the market will also have a psychological impact on the customer's quality perception of the refurbished products. In the illustrative example, we form three scenarios for customer's quality perception for the refurbished products based on the new product quality levels – Low [0.1, 0.4], Medium [0.3, 0.7], High [0.4, 0.8].

Low-end customers' acceptance rate also depends on the discount rate offered for the refurbished products. Initially, we fix the discount rate at 0.25 (25%) in all scenarios. We then vary the discount rate and determine the best discount for the illustrative example.

Given the discount rate as 25%, $rp/np = 0.75$. The low-end customer acceptance rate is calculated using the Equation (3.1). The perceived value range of $[v(q - i), v(q + i)]$ and the quality perception range for refurbished products of $[q - i, q + i]$ is referred from the hypothetical data presented in Section 3.4 of Chapter 3. We assume that the acceptance rates of low-end and high-end customers are zero when the quality level of the new product is low (65%). The acceptance rates of low-end customers are 0.3103 (31.03%) for medium quality level and 0.5385 for high-quality level. The acceptance rate for high-end customers depends only on the discount rate and is calculated using Equation (3.2). High-end customer acceptance rates are 0.1364 (13.64%) for medium and high-quality levels, at a 25% discount rate.

5.3.2 Scenarios Used in the Illustrative Example

We generate six different scenarios for the illustrative example. Based on the market demands, Scenarios 1 and 5 represent Strong demand, Scenarios 2 and

6 represent Good demand, Scenario 3 represents Fair demand, and Scenario 4 represents Weak demand. In practice, the demand is also an indicator of the quality of the new product in the market and a Strong demand also reflects that the new product quality is high and the customers are willing to buy more products; on the other hand, a Weak demand may be due to low quality. In this example, we set the new product quality High in Scenarios 1 and 6, Medium in Scenarios 2 and 5, and Low in Scenarios 3 and 4. We have three scenarios for the acceptance rates of customers based on the quality level of the new product.

The probability of occurrence of each scenario has to be determined based on the occurrence probabilities of each of the uncertain parameters. If there is only one uncertain parameter and n scenarios are defined over that uncertain parameter, we can define the probability of occurrence of each scenario based on the probability of occurrence of that single uncertain parameter. Assume a case when only the demand is uncertain, and the rest of the parameters are deterministic. Then, the illustrative example will have four scenarios with probability of occurrences based on demand. With more than one uncertain parameter, we could use a decision tree analysis to compute the probability of each scenario.

In this illustrative example, demand, return rate, and quality perception range for refurbished products are factors that go into the assessment of the occurrence probabilities of the scenarios. Since these uncertain parameters depend on a multitude of factors requiring input from several sources in the industry, the Delphi method could be used to determine the occurrence probabilities. The hypothetical data used for generating the probability of occurrence of each scenario in the illustrative example is shown in Figure 5.1. The probabilities of occurrence associated with the scenarios are: 0.20 (Scenario 1), 0.25 (Scenario 2), 0.15 (Scenario 3), 0.10 (Scenario 4), 0.15 (Scenario 5), 0.15 (Scenario 6). It is to be noted that the highlighted data associated with Scenario 2 is used in solving the deterministic model. Also, notice that this scenario has the highest probability of occurrence of 0.25 (Figure 5.1). Figure 5.1 also summarizes the three different uncertain parameters considered in the robust model, levels used for each parameter, and the final six different scenarios derived from the uncertain parameters.

Discount rate is fixed as 0.25 (25%) in all four scenarios. Acceptance rate for high-end customers is calculated using Equation (3.2) in Scenarios 1, 2, 3, 5, and, 6. Since the acceptance rate of low-end customers is zero in Scenario 4, we set the acceptance rate for high-end customers also as zero.

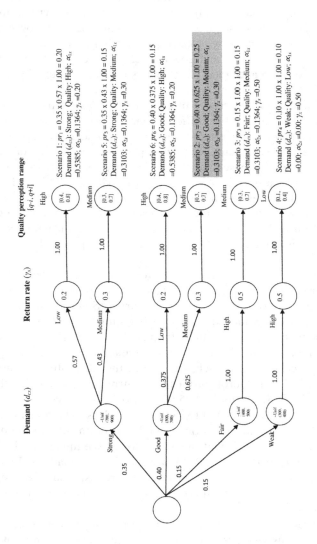

FIGURE 5.1 Decision tree analysis (probability of occurrence) and scenarios.

5.3.3 Robust Optimization Solution

The weight associated with the variance is arbitrarily set to 1, to solve the robust model. The data for the illustrative example is coded and generated in Microsoft Visual C++ 6.0. The mathematical model is coded and solved using a commercial optimization software package. The model for this example has 70,546 variables (70,466 continuous variables and 80 binary variables) and 4,736 constraints. The model took approximately 134 seconds to solve for the optimal solution.

We will first discuss the total demand, quantity of returned products, and refurbishing potential based on the customer acceptance rates in each scenario. Total demand in six scenarios are as follows: 80,028, 60,028, 44,990, 34,990, 80,028, and 60,028 units, respectively. The quantity of returns in the six scenarios are 15,823, 17,837, 22,314, 17,314, 23,837, and 11,823, respectively. Based on the customer acceptance rates, demand for refurbished products in the six scenarios are 33,391, 15,449, 11,562, 0, 20,609, and 25,031 respectively.

In Scenarios 1 and 6, customers willing to buy refurbished products are higher than the returns. In Scenarios 2, 3, and 5, the acceptance rate is lower, hence fewer customers are willing to buy refurbished products than the returns and thus some products are disposed-off at the retailers. In Scenario 4, the customer acceptance rate is zero and thus all the returned products should be disposed-off. The optimal objective function value (Max Z) achieved by the model for this example is $9,383,127, which results in an expected profit of $11,353,585. The optimal network design is as follows: Suppliers 1, 3, 5, 6, 7, 8, 9 are selected in all the six scenarios to supply raw materials to the plants. These suppliers are the primary suppliers for the raw materials. There is no variation in quantities allocated to Suppliers 1, 5, 6, 8, 9 across the scenarios and some variation for Suppliers 3 and 7. Suppliers 4 and 10 are used in the Scenarios 1, 2, 5, and 6. But, they are not used in Scenarios 3 and 4, where the demands are Fair and Weak. Suppliers 15, 17, and 19 are used only where the demand is Strong. These suppliers could be contracted as back-up suppliers, when the demand is high. In Scenarios 1, 5, and 6, all the returned units are not refurbished, even though there is demand for the refurbished products. Some of the returned products are disposed-off. This happens because, in these scenarios, it is not profitable to recycle all the return products in the supply chain. The quantities disposed-off (in each tier) in the Scenarios 1, 5, and 6 are: Scenario 1 – 4,755 (Tier 2), 1,556 (Tier 3), 1,556 (Tier 4); Scenario 5 – 7,163 (Tier 2), 2,358 (Tier 3), 2,358 (Tier 4); Scenario 6 – 1,556 (Tier 4). We observe that, in all the three

Scenarios (1, 5, and 6), capacity is available for producing new products/ refurbishing of returned products. But, we observe that the hybrid facilities stage causes the capacity bottleneck in the supply chain. The capacity utilization of the hybrid facilities in Scenarios 1 and 5 is close to 100%. Any further increase in returns or new product production might require opening a larger facility or an additional facility at some additional cost. In Scenario 6, capacity is available at the Hybrid facility 3 (its utilization is 67%). However, 1,556 of Tier 4 returns are disposed-off because it is not profitable to refurbish them that is, the sum of the cost of taking the returned products back to manufacturer, refurbishing cost, and cost of distributing the refurbished product to the retailers is higher than the cost of disposing them off. No warehouses and recovery centers are used for distribution in the optimal solution. Hybrid facilities 2, 3, and 7, each of size 3, are opened. We see that the capacity utilization of all the opened hybrid facilities in Scenarios 1 and 5, where the demand is Strong, is close to 100%. The utilization is the lowest in Scenario 4 where the demand is Weak and product quality is low. The retailers' demands are satisfied from Hybrid facilities 2, 3, and 7 in the forward channel and the returns at the retailers are shipped to Hybrid facilities 2, 3, and 7 in the return channel. The profit obtained and the demand fulfillment rate in each scenario, respectively, are shown within parenthesis: ($13,474,272, 90.6%), ($12,230,934, 100%), ($9,161,288, 100%), ($4,828,866, 100%), ($13,631,948, 85.6%), ($11,327,499, 100%). The *demand fulfillment rate* in Scenario x is the percentage of demand satisfied out of the total available demand. As we discussed earlier, the hybrid facilities cause the capacity bottlenecks in the supply chain. In particular, the demand fulfillment rate is lesser in Scenario 1 (91%) than that in Scenario 5 (86%) because of the capacity restriction in the hybrid facilities in these scenarios. In order to improve the demand fulfillment rate, more capacity is needed at the hybrid facilities at some additional cost.

5.4 ANALYSIS OF THE ROBUST OPTIMIZATION SOLUTION

In this section, we compare the deterministic model (based on the high probability Scenario 2) and the robust optimization model solutions. We fix the network designs of the supply chain obtained from the two models. We solve the deterministic model for the realizations of the six scenarios separately. In all the cases, we set the discount rate to be 0.25. We vary this later to

TABLE 5.1   Network design solutions (Deterministic & Robust optimization model)

FACILITY	DETERMINISTIC MODEL	ROBUST OPTIMIZATION MODEL
Warehouses	None	None
Hybrid facilities	2, 6, and 8, each of size 3	2, 3, and 7, each of size 3
Recovery centers	None	None

determine the best discount rate for this example. The network design solutions from the deterministic model and the robust optimization model are shown in Table 5.1.

Hybrid facilities 2, 6, and 8, each of size 3, are selected in the deterministic model and Hybrid facilities 2, 3, and 7, each of size 3, are selected in the robust optimization model. Note that the deterministic model solution corresponds to Scenario 2 of the illustrative example. Two main objectives of the supply chain design are maximizing profit and customer responsiveness. We compare profit, demand fulfillment rate, and percentage deviation of the unmet demand in each of the scenario realizations in the deterministic and robust optimizations models. The deterministic design solution is dominated by the robust supply chain design solution as it yields both higher expected profit and responsiveness (measured in terms of demand fulfillment rate). The expected profit of the supply chain in the presence of probabilistic demand, return rate, and customer acceptance rates is nearly 2% higher than that of its deterministic counterpart. The expected demand fulfilment rate in the robust model solution is nearly 3% higher and the expected unmet demand is 32% lower than the deterministic solution. Even though the increase in profit (2%) is not very significant, it does not include the cost of lost sales resulting from the 32% of unsatisfied demand. Thus, we recommend that robust optimization should be used in order to improve both the expected profit and the responsiveness of the supply chain.

5.4.1 Analysis of the Deterministic Design Solution

This section discusses the deterministic design solutions obtained in each of the six scenarios. We use this to discuss the differences between the deterministic and robust solutions. Suppliers 1, 3, 5, 6, 7, 8, 9 are selected in all the six scenarios to supply raw materials to the plants. They are the primary suppliers for the raw materials. Suppliers 4 and 10 are used in Scenarios 1, 2, 5, and 6. Supplier 15 is used only in Scenario 1. These suppliers could be contracted as back-up suppliers, when the demand is 'Strong' or 'Good'. The supplier selection results are nearly the same in the robust and the deterministic design. The total refurbished quantities in each scenario are the same in deterministic and robust models. However, the total new products produced are higher in Scenarios 1 and 5, where the demand is 'Strong'. This happens because the hybrid facility becomes the bottleneck, even though the manufacturers can produce/refurbish products. As the deterministic design is based on Scenario 2, the capacity utilization of the selected hybrid facilities is already close to 100% in Scenario 2. Hence, the deterministic supply chain design is not capable of accommodating any increase in demand or product returns (in Scenarios 1, 5, and 6).

5.4.2 Comparison of Deterministic and Robust Designs (Scenarios 1 and 2)

We now compare the deterministic and robust solutions of Scenarios 1 and 2. Scenario 1 has 20% chance of occurrence and Scenario 2 has 25% chance of occurrence (top two probable scenarios). In Scenario 1, the total supply chain profit in the robust model solution is 5.6% higher than that of the deterministic model solution. Also, the unmet demand in the robust model solution is 36.6% lower than the deterministic model. In Scenario 2, the profit in the robust model solution is 0.5% lesser than that of the deterministic model solution, which is not significant. When the deterministic solution is implemented in the presence of probabilistic demand, the solution might be sub-optimal. Particularly, in our analysis the robust solution is more effective when demand is Strong (higher than the demand used in designing the deterministic model). The robust model design takes into account the extreme cases in the scenarios.

5.5 ADVANTAGES OF ROBUST OPTIMIZATION

Scenario 2 has the highest probability of occurrence (25%) among all the scenarios and is used to build the deterministic design solution. In this section, we will generate two scenarios based on modifications to

Scenario 2 as follows: (1) demand: 'Good', product return rate: 0.50, quality perception range of refurbished products: 'High': in this scenario, we increase the fraction of product returns, and customer acceptance rate for refurbished products. This is a new scenario and is not one of the six scenarios in the illustrative example; (2) demand: 'Strong', product return rate: 0.30, quality perception range of refurbished products: 'Medium': in this scenario, we increase the market demand keeping the fraction of product returns, and customer acceptance rate for refurbished products same as that of Scenario 2. This is same as Scenario 5 in the illustrative example.

The objective of this section is to study the case where there is a slight modification to the most probable scenario.

5.5.1 Scenario with 'Good' Demand – Increasing Return Rate and Acceptance Rate

Consider Scenario 2 where the demand is 'Good', with low-end customer acceptance rate of 0.3103, high-end customer acceptance rate of 0.1364, and return rate of 0.30. Note that Scenario 2 has the highest probability of occurrence. Let us assume a case where the return rate increases to 0.50 due to higher customer remorse, and the customer acceptance rate also increases due to advertising and promotional campaigns by the company (low-end customer acceptance rate = 0.5385, high-end customer acceptance rate = 0.1364). The demand still remains the same, that is, 'Good' ~unif(500, 700). As this is a new scenario, we solve the network design model with the robust design and the deterministic design solutions (refer to Table 5.1 for the robust and deterministic design solutions).

The total demand in this scenario is 60,028 units and the total returns are 29,837 units. The total customer acceptance rate is 25,031. The production and refurbishing plans at the plants are as follows: 43,968 units of new products and 16,060 units of refurbished products are produced in the deterministic solution; 39,618 units of new products and 20,410 units of refurbished products are produced in the robust solution. The robust design solution is capable of accommodating higher returns than the deterministic design solution, as we can see that it refurbished more. The deterministic solution produces more new products to satisfy the customer demands. The total supply chain profit from the robust design and deterministic design are $13,501,981 and $12,945,963, respectively. The robust design yields a profit of 4.3% higher than the deterministic design.

5.5.2 Scenario with Medium Return Rate and Acceptance Rate – Increasing the Demand

Now, consider Scenario 2 and assume a case where the acceptance rates and return rate do not change but the demand becomes 'Strong' ~unif(700, 500), that is, Scenario 5. The robust design solution gives 5.6% higher profit, 6.4% higher demand fulfillment rate, and 27.4% less unmet demand than its deterministic counterpart.

5.6 OPTIMAL DISCOUNT FOR THE ROBUST MODEL

In this section, we study the effect of varying discounts offered for the refurbished products on the total profit and design of the supply chain network. In this analysis, we fix the mix of low-end and high-end customers at 70% and 30%, respectively. To determine the optimal discount rate for the robust and the deterministic models, we perform the following analysis separately:

Optimal discount for robust model: We will vary the discounts offered for refurbished products between 0.04 and 0.54 in increments of 0.02 (26 levels). A total of 26 cases have been modeled and optimal solutions are obtained. We consider the illustrative example with all the six scenarios.

Optimal discount for deterministic model: The best discount rate for the deterministic model is determined using only Scenario 2. We vary the discounts offered for refurbished products between 0.04 and 0.54 in increments of 0.02.

The product quality in the market and the corresponding perceived value for the refurbished products vary across the scenarios. The customer acceptance rate for low-end customers also depends on the discount offered for the refurbished products. Thus, we use the equations based on Equation (3.1) (see Chapter 3), to calculate the customer acceptance rate for low-end customers. For Scenarios 1 and 6, the low-end customer acceptance rate is 0% for discounts up to 11%. Hence, we set the customer acceptance rate for high-end customers also at 0% for discounts up to 11%. For discount levels higher than 11%, the customer acceptance rate for high-end customers is calculated using Equation (3.2) (see Chapter 3). For Scenarios 2, 3, and 5, as the low-end customer acceptance rate is 0% for discounts up to 16%, we set the customer acceptance rate for high-end customers to be also at 0% up to 16% discount level. For discount levels higher than 16%, the customer acceptance rate for high-end customers is calculated using Equation (3.2). For Scenario 4, the high-end customer acceptance rate is 0. The expected profit of the supply chain in the robust model and the profit of the supply chain in the deterministic model under different levels of discount offered for refurbished products are computed.

Optimal discount for robust model: We observe that the expected profit of the robust model also follows an inverted U-shaped curve. The discounts offered have a significant impact on the total supply chain profit. The analysis shows that the maximum expected profit occurs at a discount of 25%.

Optimal discount for deterministic model: It should be noted that the deterministic model is run only using the Scenario 2 of the illustrative example. The maximum profit for the deterministic design occurs at a discount of 24%. Note that the expected profit of the robust model is based on all the six scenarios. The maximum profit of the deterministic model is only based on Scenario 2. Since the deterministic design is optimized for a specific scenario realization, it has a higher expected profit.

5.6.1 Comparison between the Deterministic and Robust Designs (Optimal Discount)

Considering only the deterministic data, management generally tends to use 24% as the discount rate for refurbished products. We will now solve the deterministic model for the realizations of the six scenarios separately using the network design shown in Table 5.2 (deterministic design solution for 24% discount rate). We already have the solution for the robust design solution in the illustrative example section (robust design solution for 25% discount rate).

TABLE 5.2 Network design solution for the optimal discount rate (Deterministic model)

FACILITY	DETERMINISTIC MODEL
Warehouses	11 of size 3
Hybrid facilities	6, and 8, each of size 3
Recovery centers	None

We will begin with analyzing the profit and the demand fulfillment rate in each of the scenario realizations under the optimal network design and discount levels obtained from the deterministic and robust models. Note that the optimal discount rates are 25% for robust design and 24% for deterministic design. The profit and demand fulfillment rate in each of the scenario realizations in the deterministic and robust optimizations solutions using their optimal discount levels, i.e., we use 25% discount rate for robust design and 24% discount rate for deterministic design, are analyzed.

The expected profit of the supply chain in the presence of probabilistic demand, return rate, and customer acceptance rates is 3% higher than that of its deterministic counterpart. The expected demand fulfillment rate is 4% higher and the unmet demand is 42% lower than the deterministic solution. In Scenario 1 (with 20% probability of occurrence), the robust solution yields 8.1% higher profit and 9.1% higher demand fulfillment rate than the deterministic solution, which is significant. Thus, the discount rates should be set considering uncertainty in the data.

In summary, the following are the inferences from the robust optimization analysis: (1) the expected profit and the demand fulfillment rate of the robust model are higher than that of the deterministic model solutions. Also, the unmet demand in the robust design is less than that of the deterministic design; (2) the robust model is more efficient in accommodating higher demand in the supply chain and when the returns are higher; (3) in the presence of uncertainty in demand and return parameters, the robust model is more effective when compared to the deterministic model; (4) the discount levels have a significant impact on the expected profit and customer responsiveness of the supply chain. The discount rate has to be set considering the uncertain parameters in order to avoid sub-optimal solutions.

5.7 SUMMARY

In this chapter, we extend the deterministic model presented in Chapter 4 to consider uncertainty in demand, return rate, and customer acceptance rates. This chapter addresses the question of how to incorporate uncertainties of input parameters in the supply chain design phase. The framework allows companies to incorporate any number of uncertain parameters while designing the CLSC networks. These uncertain parameters are modeled as a set of discrete future scenarios and are associated with a probability of occurrence, determined by a decision maker. We develop a scenario-based robust optimization model for the problem. The objective function of the model was to maximize the expected profit of the supply chain along with minimizing the variability of the profits across the scenarios. Minimizing the variability is important for mitigating risk associated with one or more uncertain input parameters in the supply chain.

We illustrate the model using an example with six scenarios. The output from the robust model is comprehensive and can bring valuable insights during design phases. For example, the reason for allowing unmet demand in certain scenarios can be explained using model outputs. We provide a comparison between the robust design solution and the deterministic design solution. The analysis shows that the robust design yields 2% higher expected profit, 3% higher expected demand fulfillment rate, and 32% less expected unmet demand than the deterministic design. We show that the robust design solution is more effective when there is an increase in demand and/or the product return quantities.

The analysis in this chapter also shows that the discount rate should be set considering the uncertain parameters, in order to avoid sub-optimal solutions. The analysis showed that the robust model yields nearly 3% higher expected profit, 4% higher demand fulfillment rate, and 42% less expected unmet demand compared to its deterministic counterpart. This emphasizes the importance of considering uncertainty in the input parameters and using robust optimization models to design the CLSC network, in order to avoid the loss of potential profits and customer responsiveness.

Similar to other models, it can be challenging to collect data on customer behavior and return rates. But, modeling these parameters incorporating uncertainty significantly mitigates the risk of operating in a sub-optimal supply chain design. This model can also be extended using multiple objectives depending on the requirements of the companies.

REFERENCES

Alonso-Ayuso, A., Escudero, L. F., Garin, A., Ortuno, M. T., Perez, G. (2003), 'An approach for strategic supply chain planning under uncertainty based on stochastic 0-1 programming', *Journal of Global Optimization* **26**(1), 97–124.

Charnes, A., Cooper, W. W., Symonds, G. H. (1958), 'Cost horizons and certainty equivalents: An approach to stochastic programming of heating oil', *Management Science* **4**(3), 235–263.

Dantzig, G. B. (1955), 'Linear programming under uncertainty', *Management Science* **1**(3/4), 197–206.

Greenberg, H. J., Morrison, T. (2008), 'Robust Optimization', In *Operations Research and Management Science Handbook*, ed. A. Ravi Ravindran, Chapter 14. Boca Raton, FL, CRC Press

Guillén, G., Mele, F., Bagajewicz, M., Espuna, A., Puigjaner, L. (2005), 'Multiobjective supply chain design under uncertainty', *Chemical Engineering Science* **60**(6), 1535–1553.

Gupta, A., Maranas, C. (2003), 'Managing demand uncertainty in supply chain planning', *Computers and Chemical Engineering* **27**(8–9), 1219–1227.

Gupta, S. K., Rosenhead, J. (1968), 'Robustness in sequential investment decisions', *Management Science* **15**(2), B–18–29.

Leung, S. C. H., Wu, Y., Lai, K. K. (2006), 'A stochastic programming approach for multi-site aggregate production planning', *Journal of the Operational Research Society* **57**, 123–132.

Markowitz, H. (1952), 'Portfolio selection', *Journal of Finance* **7**(1), 77–91.

Mulvey, J. M., Vanderbei, R. J., Zenios, S. A. (1995), 'Robust optimization of large-scale systems', *Operations Research* **43**(2), 264–281.

Pazhani (2014), 'Design of Closed-Loop Supply Chain Networks', The Pennsylvania State University, https://etda.libraries.psu.edu/files/final_submissions/9801

Ravindran, A. R. (2008), *Operations Research and Management Science Handbook*, CRC Press, Boca Raton, FL.

Santoso, T., Ahmed, S., Goetschalckx, M., Shapiro, A. (2005), 'A stochastic programming approach for supply chain network design under uncertainty', *European Journal of Operational Research* **167**(1), 96–115.

Sen, S., Higle, J. (1999), 'An introductory tutorial on stochastic linear programming', *Interfaces* **29**(2), 33–61.

Solo, C. J. (2009), 'Multi-Objective, Integrated Supply Chain Design and Operation Under Uncertainty', PhD thesis, The Pennsylvania State University, University Park, PA.

Yu, C. S., Li, H. L. (2000), 'A robust optimization model for stochastic logistic problems', *International Journal of Production Economics* **64**(1–3), 385–397.

Printed in the United States
by Baker & Taylor Publisher Services